Introduction to Bacteriology

Introduction to Bacteriology

Edited by
Haris Russell

Larsen & Keller
www.larsen-keller.com

Introduction to Bacteriology
Edited by Haris Russell
ISBN: 978-1-63549-037-4 (Hardback)

☰ Larsen & Keller

Published by Larsen and Keller Education,
5 Penn Plaza,
19th Floor,
New York, NY 10001, USA

Cataloging-in-Publication Data

Introduction to bacteriology / edited by Haris Russell.
    p. cm.
Includes bibliographical references and index.
ISBN 978-1-63549-037-4
1. Bacteriology. 2. Bacteria. 3. Bacterial diseases. I. Russell, Haris.
QR74.8 .I58 2017
579.3--dc23

The publisher's policy is to use permanent paper from mills that operate a sustainable forestry policy. Furthermore, the publisher ensures that the text paper and cover boards used have met acceptable environmental accreditation standards.

Printed and bound in the United States of America.

For more information regarding Larsen and Keller Education and its products, please visit the publisher's website www.larsen-keller.com

Table of Contents

Permissions

Index

Preface

This book elucidates the concepts and innovative models around prospective developments with respect to bacteriology. It provides in depth information about the field and its applications. Bacteriology is a part of microbiology. It refers to the study of the classification, identification and characterization of bacteria which is a prokaryotic microorganism. This text will give knowledge about the uses of bacteria in the various industries and their importance in medicinal studies. Most of the topics introduced in the book cover new techniques and the applications of bacteriology. Through this book, we attempt to further enlighten the readers about the new concepts in this field.

To facilitate a deeper understanding of the contents of this book a short introduction of every chapter is written below:

Chapter 1- The study of bacteria is termed as bacteriology. A bacteriologist is a person who studies bacteriology. Bacteria can be lethal as well as benign. This chapter on bacteriology offers an insightful focus, keeping in mind the complex subject matter.

Chapter 2- Bacterial taxonomy is the classification of bacteria on the basis of their rank in the taxonomic hierarchy. Some of the examples of these ranks are genus, family, kingdom etc. Topics such as monera and bacterial phyla have also been elucidated in the following section. The topics discussed in the chapter are of great importance to broaden the existing knowledge on bacteriology.

Chapter 3- The types of bacteria that have been discussed in the chapter are epiphytic bacteria, gram-negative bacteria, gram-positive bacteria, indicator bacteria and cyanobacteria. Epiphytic bacteria live on the surface of a plant; different bacterias prefer different plants to live on. This section explains to the reader the types of bacteria that exist.

Chapter 4- In order to study proteins in bacteria it is necessary to understand AB5 toxin, actin assembly-inducing protein, bacterial effector protein, cholera toxin and lac repressor. AB5 toxins have six component protein complexes and bacterial effectors are proteins secreted by pathogenic bacteria. This chapter serves as a source to understand bacteria in an explicated manner.

Chapter 5- Bacteria can cause a number of diseases and some of these diseases are bacterial pneumonia, ehrlichia ruminantium, actinomycosis, leprosy and tuberculosis. Bacterial pneumonia is a particular type of pneumonia caused by bacteria and leprosy is also a disease caused by bacteria but the symptoms of this disease usually go unnoticed for initial couple of years. This chapter ha been carefully written to provide an easy understanding of the diseases caused by bacteria.

I owe the completion of this book to the never-ending support of my family, who supported me throughout the project.

Editor

Introduction to Bacteriology

The study of bacteria is termed as bacteriology. A bacteriologist is a person who studies bacteriology. Bacteria can be lethal as well as benign. This chapter on bacteriology offers an insightful focus, keeping in mind the complex subject matter.

Bacteriology

Bacteriology is the study of bacteria. This subdivision of microbiology involves the identification, classification, and characterization of bacterial species. A person who studies bacteriology is a bacteriologist.

Bacteriology and Microbiology

Because of the similarity of thinking and working with microorganisms other than bacteria, such as protozoa, fungi, and viruses, there has been a tendency for the field of bacteriology to extend as microbiology. The terms were formerly often used interchangeably. However, bacteriology can be classified as a distinct science.

Bacteria

Bacteria (common noun bacteria, singular bacterium) constitute a large domain of prokaryotic microorganisms. Typically a few micrometres in length, bacteria have a number of shapes, ranging from spheres to rods and spirals. Bacteria were among the first life forms to appear on Earth, and are present in most of its habitats. Bacteria inhabit soil, water, acidic hot springs, radioactive waste, and the deep portions of Earth's crust. Bacteria also live in symbiotic and parasitic relationships with plants and animals.

There are typically 40 million bacterial cells in a gram of soil and a million bacterial cells in a millilitre of fresh water. There are approximately 5×10^{30} bacteria on Earth, forming a biomass which exceeds that of all plants and animals. Bacteria are vital in recycling nutrients, with many of the stages in nutrient cycles dependent on these organisms, such as the fixation of nitrogen from the atmosphere and putrefaction. In the biological communities surrounding hydrothermal vents and cold seeps, bacteria provide the nutrients needed to sustain life by converting dissolved compounds, such as hydrogen sulphide and methane, to energy. On 17 March 2013, researchers reported data that suggested bacterial life forms thrive in the Mariana Trench, which with a depth of up to 11 kilometres is the deepest part of the Earth's oceans. Other researchers reported related studies that microbes thrive inside rocks up to 580 metres below the sea floor under 2.6 kilometres of ocean off the coast of the northwestern United States. According to one of the researchers, "You can find microbes everywhere — they're extremely adaptable to conditions, and survive wherever they are."

Most bacteria have not been characterised, and only about half of the bacterial phyla have species that can be grown in the laboratory. The study of bacteria is known as bacteriology, a branch of microbiology.

There are approximately ten times as many bacterial cells in the human flora as there are human cells in the body, with the largest number of the human flora being in the gut flora, and a large number on the skin. The vast majority of the bacteria in the body are rendered harmless by the protective effects of the immune system, and some are beneficial. However, several species of bacteria are pathogenic and cause infectious diseases, including cholera, syphilis, anthrax, leprosy, and bubonic plague. The most common fatal bacterial diseases are respiratory infections, with tuberculosis alone killing about 2 million people per year, mostly in sub-Saharan Africa. In developed countries, antibiotics are used to treat bacterial infections and are also used in farming, making antibiotic resistance a growing problem. In industry, bacteria are important in sewage treatment and the breakdown of oil spills, the production of cheese and yogurt through fermentation, and the recovery of gold, palladium, copper and other metals in the mining sector, as well as in biotechnology, and the manufacture of antibiotics and other chemicals.

Once regarded as plants constituting the class *Schizomycetes*, bacteria are now classified as prokaryotes. Unlike cells of animals and other eukaryotes, bacterial cells do not contain a nucleus and rarely harbour membrane-bound organelles. Although the term *bacteria* traditionally included all prokaryotes, the scientific classification changed after the discovery in the 1990s that prokaryotes consist of two very different groups of organisms that evolved from an ancient common ancestor. These evolutionary domains are called *Bacteria* and *Archaea*.

Etymology

The word *bacteria* is the plural of the New Latin *bacterium*, which is the latinisation of the Greek βακτήριον (*bakterion*), the diminutive of βακτηρία (*bakteria*), meaning "staff, cane", because the first ones to be discovered were rod-shaped.

Origin and Early Evolution

The ancestors of modern bacteria were unicellular microorganisms that were the first forms of life to appear on Earth, about 4 billion years ago. For about 3 billion years, most organisms were microscopic, and bacteria and archaea were the dominant forms of life. In 2008, fossils of macroorganisms were discovered and named as the Francevillian biota. Although bacterial fossils exist, such as stromatolites, their lack of distinctive morphology prevents them from being used to examine the history of bacterial evolution, or to date the time of origin of a particular bacterial species. However, gene sequences can be used to reconstruct the bacterial phylogeny, and these studies indicate that bacteria diverged first from the archaeal/eukaryotic lineage. Bacteria were also involved in the second great evolutionary divergence, that of the archaea and eukaryotes. Here, eukaryotes resulted from the entering of ancient bacteria into endosymbiotic associations with the ancestors of eukaryotic cells, which were themselves possibly related to the Archaea. This involved the engulfment by proto-eukaryotic cells of alphaproteobacterial symbionts to form either mitochondria or hydrogenosomes, which are still found in all known Eukarya (sometimes in highly reduced form, e.g. in ancient "amitochondrial" protozoa). Later on, some eukaryotes that already contained mitochondria also engulfed cyanobacterial-like organisms. This led to the formation of chloroplasts in algae and plants. There are also some algae that originated from even later endosymbiotic events. Here, eukaryotes engulfed a eukaryotic algae that developed into a "second-generation" plastid. This is known as secondary endosymbiosis.

Morphology

Bacteria display a wide diversity of shapes and sizes, called *morphologies*. Bacterial cells are about one-tenth the size of eukaryotic cells and are typically 0.5–5.0 micrometres in length. However, a few species are visible to the unaided eye — for example, *Thiomargarita namibiensis* is up to half a millimetre long and *Epulopiscium fishelsoni* reaches 0.7 mm. Among the smallest bacteria are members of the genus *Mycoplasma*, which measure only 0.3 micrometres, as small as the largest viruses. Some bacteria may be even smaller, but these ultramicrobacteria are not well-studied.

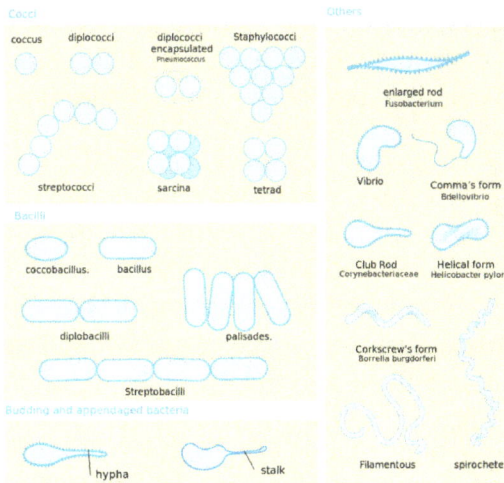

Bacteria display many cell morphologies and arrangements

Most bacterial species are either spherical, called *cocci*, or rod-shaped, called *bacilli*. Elongation is associated with swimming. Some bacteria, called *vibrio*, are shaped like slightly curved rods or comma-shaped; others can be spiral-shaped, called *spirilla*, or tightly coiled, called *spirochaetes*. A small number of species even have tetrahedral or cuboidal shapes. More recently, some bacteria were discovered deep under Earth's crust that grow as branching filamentous types with a star-shaped cross-section. The large surface area to volume ratio of this morphology may give these bacteria an ad-vantage in nutrient-poor environments. This wide variety of shapes is determined by the bacterial cell wall and cytoskeleton, and is important because it can influence the ability of bacteria to acquire nutrients, attach to surfaces, swim through liquids and escape predators.

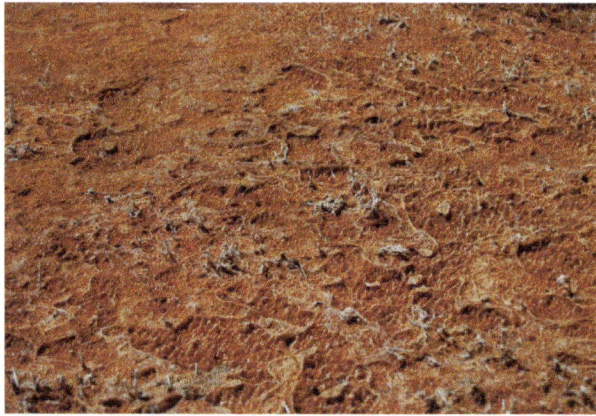

A biofilm of thermophilic bacteria in the outflow of Mickey Hot Springs,
Oregon, approximately 20 mm thick.

Many bacterial species exist simply as single cells, others associate in characteristic patterns: *Neisseria* form diploids (pairs), *Streptococcus* form chains, and *Staphylococcus* group together in "bunch of grapes" clusters. Bacteria can also be elongated to form filaments, for example the Actinobacteria. Filamentous bacteria are often surrounded by a sheath that contains many individual cells. Certain types, such as species of the genus *Nocardia*, even form complex, branched filaments, similar in appearance to fungal mycelia.

Bacteria often attach to surfaces and form dense aggregations called *biofilms* or bacterial mats. These films can range from a few micrometres in thickness to up to half a metre in depth, and may contain multiple species of bacteria, protists and archaea. Bacteria living in biofilms display a complex arrangement of cells and extracellular components, forming secondary structures, such as microcolonies, through which there are networks of channels to enable better diffusion of nutrients. In natural environments, such as soil or the surfaces of plants, the majority of bacteria are bound to surfaces in biofilms. Biofilms are also important in medicine, as these structures are often present during chronic bacterial infections or in infections of implanted medical devices, and bacteria protected within biofilms are much harder to kill than individual isolated bacteria.

Even more complex morphological changes are sometimes possible. For example, when starved of amino acids, Myxobacteria detect surrounding cells in a process known as quorum sensing, migrate towards each other, and aggregate to form fruiting bodies up to 500 micrometres long and containing approximately 100,000 bacterial cells. In these fruiting bodies, the bacteria perform separate tasks; this type of cooperation is a simple type of multicellular organisation. For example, about one in 10 cells migrate to the top of these fruiting bodies and differentiate into a specialised dormant state called myxospores, which are more resistant to drying and other adverse environmental conditions than are ordinary cells.

Cellular Structure

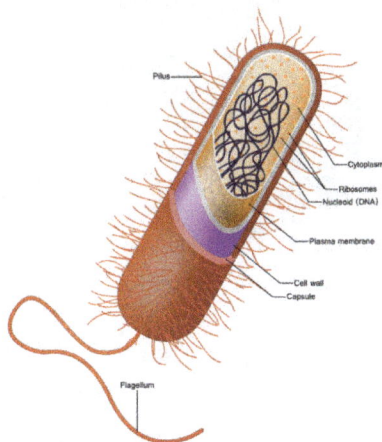

Structure and contents of a typical gram-positive bacterial cell (seen by the fact that only *one* cell membrane is present).

Intracellular Structures

The bacterial cell is surrounded by a cell membrane (also known as a lipid, cytoplasmic or plasma membrane). This membrane encloses the contents of the cell and acts as a barrier to hold nutrients, proteins and other essential components of the *cytoplasm* within the cell. As they are prokaryotes, bacteria do not usually have membrane-bound organelles in their cytoplasm, and thus contain few large intracellular structures. They lack a true nucleus, mitochondria, chloroplasts and the other organelles present in eukaryotic cells. Bacteria were once seen as simple bags of cytoplasm, but structures such as the *prokaryotic cytoskeleton* and the localisation of proteins to specific locations within the cytoplasm that give bacteria some complexity have been discovered. These subcellular levels of organisation have been called "bacterial hyperstructures".

Bacterial microcompartments, such as carboxysomes, provide a further level of organisation; they are compartments within bacteria that are surrounded by polyhedral protein shells, rather than by lipid membranes. These "polyhedral organelles" localise and compartmentalise bacterial metabolism, a function performed by the membrane-bound organelles in eukaryotes.

Many important biochemical reactions, such as energy generation, use concentration gradients across membranes. The general lack of internal membranes in bacteria means reactions such as electron transport occur across the cell membrane between the cytoplasm and the *periplasmic space*. However, in many photosynthetic bacteria the plasma membrane is highly folded and fills most of the cell with layers of light-gathering membrane. These light-gathering complexes may even form lipid-enclosed structures called chlorosomes in green sulfur bacteria. Other proteins import nutrients across the cell membrane, or expel undesired molecules from the cytoplasm.

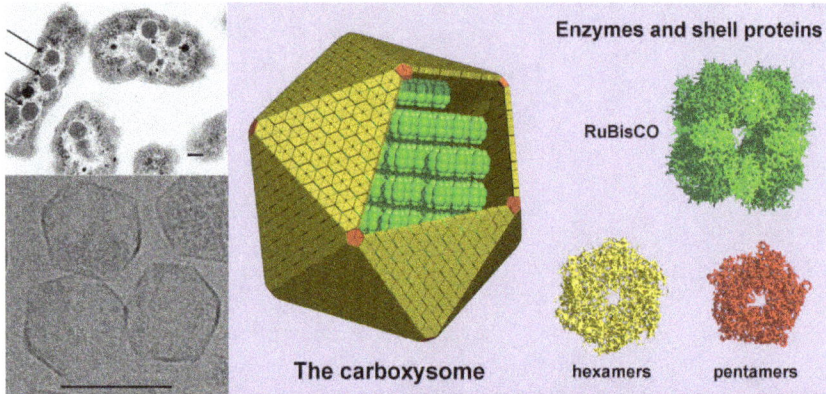

Carboxysomes are protein-enclosed bacterial organelles. Top left is an electron microscope image of carboxysomes in *Halothiobacillus neapolitanus*, below is an image of purified carboxysomes. On the right is a model of their structure. Scale bars are 100 nm.

Bacteria do not have a membrane-bound nucleus, and their genetic material is typically a single circular DNA chromosome located in the cytoplasm in an irregularly shaped body called the *nucleoid*. The nucleoid contains the chromosome with its associated proteins and RNA. The phylum Planctomycetes and candidate phylum Poribacteria may be exceptions to the general absence of internal membranes in bacteria, because they appear to have a double membrane around their nucleoids and contain other membrane-bound cellular structures. Like all living organisms, bacteria contain *ribosomes*, often grouped in chains called polyribosomes, for the production of proteins, but the structure of the bacterial ribosome is different from that of eukaryotes and Archaea. Bacterial ribosomes have a sedimentation rate of 70S (measured in Svedberg units): their subunits have rates of 30S and 50S. Some antibiotics bind specifically to 70S ribosomes and inhibit bacterial protein synthesis. Those antibiotics kill bacteria without affecting the larger 80S ribosomes of eukaryotic cells and without harming the host.

Some bacteria produce intracellular nutrient storage granules for later use, such as glycogen, polyphosphate, sulfur or polyhydroxyalkanoates. Certain bacterial species, such as the photosynthetic Cyanobacteria, produce internal gas vesicles, which they use to regulate their buoyancy – allowing them to move up or down into water layers with different light intensities and nutrient levels. *Intracellular membranes* called *chromatophores* are also found in membranes of phototrophic bacteria. Used primarily for photosynthesis, they contain bacteriochlorophyll pigments and carotenoids. An early idea was that bacteria might contain membrane folds termed mesosomes, but these were later shown to be artefacts produced by the chemicals used to prepare the cells for electron microscopy. *Inclusions* are considered to be nonliving components of the cell that do not possess metabolic activity and are not bounded by membranes. The most common inclusions are glycogen, lipid droplets, crystals, and pigments. *Volutin granules* are cytoplasmic inclusions of complexed inorganic polyphosphate. These granules are called *metachromatic granules* due to their displaying the metachromatic effect; they appear red or blue when stained with the blue dyes methylene blue or toluidine blue. *Gas vacuoles*, which are freely permeable to gas, are membrane-bound vesicles present in some species of *Cyanobacteria*. They allow the bacteria to control their buoyancy. *Microcompartments* are widespread, membrane-bound organelles that are made of a protein shell that surrounds and encloses various enzymes. *Carboxysomes* are bacterial microcompartments that contain enzymes involved in carbon fixation. *Magnetosomes* are bacterial microcompartments, present in magnetotactic bacteria, that contain magnetic crystals.

Extracellular Structures

In most bacteria, a *cell wall* is present on the outside of the cell membrane. The cell membrane and cell wall comprise the *cell envelope*. A common bacterial cell wall material is *peptidoglycan* (called "murein" in older sources), which is made from polysaccharide chains cross-linked by peptides containing D-amino acids. Bacterial cell walls are different from the cell walls of plants and fungi, which are made of cellulose and chitin, respectively. The cell wall of bacteria is also distinct from that of Archaea, which do not contain peptidoglycan. The cell wall is essential to the survival of many bacteria, and the antibiotic penicillin is able to kill bacteria by inhibiting a step in the synthesis of peptidoglycan.

There are broadly speaking two different types of cell wall in bacteria, a thick one in the gram-positives and a thinner one in the gram-negatives. The names originate from the reaction of cells to the Gram stain, a long-standing test for the classification of bacterial species.

Gram-positive bacteria possess a thick cell wall containing many layers of peptidoglycan and *teichoic acids*. In contrast, *gram-negative bacteria* have a relatively thin cell wall consisting of a few layers of peptidoglycan surrounded by a second lipid membrane containing *lipopolysaccharides* and lipoproteins. Lipopolysaccharides, also called *endotoxins*, are composed of polysaccharides and *lipid A* that is responsible for much of the toxicity of gram-negative bacteria. Most bacteria have the gram-negative cell wall, and only the Firmicutes and Actinobacteria have the alternative gram-positive arrangement. These two groups were previously known as the low G+C and high G+C gram-positive bacteria, respectively. These differences in structure can produce differences in antibiotic susceptibility; for instance, vancomycin can kill only gram-positive bacteria and is ineffective against gram-negative pathogens, such as *Haemophilus influenzae* or *Pseudomonas aeruginosa*. If the bacterial cell wall is entirely removed, it is called a *protoplast*, whereas if it is partially removed, it is called a *spheroplast*. β-Lactam antibiotics, such as penicillin, inhibit the formation of peptidoglycan cross-links in the bacterial cell wall. The enzyme lysozyme, found in human tears, also digests the cell wall of bacteria and is the body's main defence against eye infections.

Acid-fast bacteria, such as *Mycobacteria*, are resistant to decolorisation by acids during staining procedures. The high mycolic acid content of *Mycobacteria*, is responsible for the staining pattern of poor absorption followed by high retention. The most common staining technique used to identify acid-fast bacteria is the Ziehl-Neelsen stain or acid-fast stain, in which the acid-fast bacilli are stained bright-red and stand out clearly against a blue background. *L-form bacteria* are strains of bacteria that lack cell walls. The main pathogenic bacteria in this class is *Mycoplasma*.

In many bacteria, an *S-layer* of rigidly arrayed protein molecules covers the outside of the cell. This layer provides chemical and physical protection for the cell surface and can act as a macromolecular diffusion barrier. S-layers have diverse but mostly poorly understood functions, but are known to act as virulence factors in *Campylobacter* and contain surface enzymes in *Bacillus stearothermophilus*.

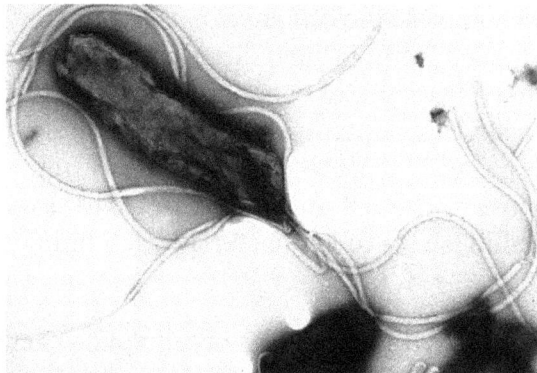

Helicobacter pylori electron micrograph, showing multiple flagella on the cell surface

Flagella are rigid protein structures, about 20 nanometres in diameter and up to 20 micrometres in length, that are used for motility. Flagella are driven by the energy released by the transfer of ions down an electrochemical gradient across the cell membrane.

Fimbriae (sometimes called "attachment pili") are fine filaments of protein, usually 2–10 nanometres in diameter and up to several micrometres in length. They are distributed over the surface of the cell, and resemble fine hairs when seen under the electron microscope. Fimbriae are believed to be involved in attachment to solid surfaces or to other cells, and are essential for the virulence of some bacterial pathogens. *Pili* (*sing.* pilus) are cellular appendages, slightly larger than fimbriae, that can transfer genetic material between bacterial cells in a process called conjugation where they are called *conjugation pili* or "sex pili". They can also generate movement where they are called *type IV pili*.

Glycocalyx are produced by many bacteria to surround their cells, and vary in structural complexity: ranging from a disorganised *slime layer* of extra-cellular polymer to a highly structured *capsule*. These structures can protect cells from engulfment by eukaryotic cells such as macrophages (part of the human immune system). They can also act as antigens and be involved in cell recognition, as well as aiding attachment to surfaces and the formation of biofilms.

The assembly of these extracellular structures is dependent on bacterial secretion systems. These transfer proteins from the cytoplasm into the periplasm or into the environment around the cell. Many types of secretion systems are known and these structures are often essential for the virulence of pathogens, so are intensively studied.

Endospores

Certain genera of gram-positive bacteria, such as *Bacillus, Clostridium, Sporohalobacter, Anaerobacter*, and *Heliobacterium*, can form highly resistant, dormant structures called *endospores*. In almost all cases, one endospore is formed and this is not a reproductive process, although *Anaerobacter* can make up to seven endospores in a single cell. Endospores have a central core of cytoplasm containing DNA and ribosomes surrounded by a cortex layer and protected by an impermeable and rigid coat. Dipicolinic acid is a chemical compound that composes 5% to 15% of the dry weight of bacterial spores. It is implicated as responsible for the heat resistance of the endospore.

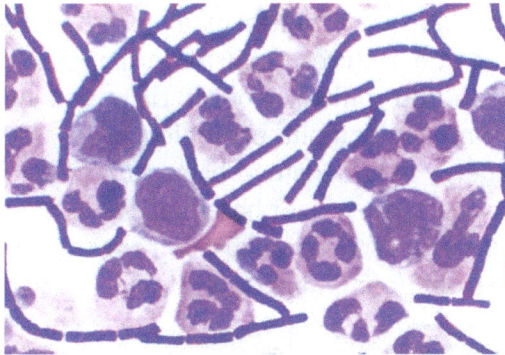

Bacillus anthracis (stained purple) growing in cerebrospinal fluid

Endospores show no detectable metabolism and can survive extreme physical and chemical stresses, such as high levels of UV light, gamma radiation, detergents, disinfectants, heat, freezing, pressure, and desiccation. In this dormant state, these organisms may remain viable for millions of years, and endospores even allow bacteria to survive exposure to the vacuum and radiation in space. According to scientist Dr. Steinn Sigurdsson, "There are viable bacterial spores that have been found that are 40 million years old on Earth — and we know they're very hardened to radiation." Endospore-forming bacteria can also cause disease: for example, anthrax can be contracted by the inhalation of *Bacillus anthracis* endospores, and contamination of deep puncture wounds with *Clostridium tetani* endospores causes tetanus.

Metabolism

Bacteria exhibit an extremely wide variety of metabolic types. The distribution of metabolic traits within a group of bacteria has traditionally been used to define their taxonomy, but these traits often do not correspond with modern genetic classifications. Bacterial metabolism is classified into nutritional groups on the basis of three major criteria: the kind of energy used for growth, the source of carbon, and the electron donors used for growth. An additional criterion of respiratory microorganisms are the electron acceptors used for aerobic or anaerobic respiration.

Nutritional types in bacterial metabolism			
Nutritional type	**Source of energy**	**Source of carbon**	**Examples**
Phototrophs	Sunlight	Organic compounds (photoheterotrophs) or carbon fixation (photoautotrophs)	Cyanobacteria, Green sulfur bacteria, Chloroflexi, or Purple bacteria
Lithotrophs	Inorganic compounds	Organic compounds (lithoheterotrophs) or carbon fixation (lithoautotrophs)	Thermodesulfobacteria, *Hydrogenophilaceae*, or Nitrospirae
Organotrophs	Organic compounds	Organic compounds (chemoheterotrophs) or carbon fixation (chemoautotrophs)	*Bacillus, Clostridium* or *Enterobacteriaceae*

Carbon metabolism in bacteria is either *heterotrophic*, where organic carbon compounds are used as carbon sources, or *autotrophic*, meaning that cellular carbon is obtained by fixing carbon dioxide. Heterotrophic bacteria include parasitic types. Typical autotrophic bacteria are phototrophic cyanobacteria, green sulfur-bacteria and some purple bacteria, but also many chemolithotrophic species, such as nitrifying or sulfur-oxidising bacteria. Energy metabolism of bacteria is either based on *phototrophy*, the use of light through photosynthesis, or based on *chemotrophy*, the use of chemical substances for energy, which are mostly oxidised at the expense of oxygen or alternative electron acceptors (aerobic/anaerobic respiration).

Filaments of photosynthetic cyanobacteria

Bacteria are further divided into *lithotrophs* that use inorganic electron donors and *organotrophs* that use organic compounds as electron donors. Chemotrophic organisms use the respective electron donors for energy conservation (by aerobic/anaerobic respiration or fermentation) and biosynthetic reactions (e.g., carbon dioxide fixation), whereas phototrophic organisms use them only for biosynthetic purposes. Respiratory organisms use chemical compounds as a source of energy by taking electrons from the reduced substrate and transferring them to a terminal electron acceptor in a redox reaction. This reaction releases energy that can be used to synthesise ATP and drive metabolism. In *aerobic organisms*, oxygen is used as the electron acceptor. In *anaerobic organisms* other inorganic compounds, such as nitrate, sulfate or carbon dioxide are used as electron acceptors. This leads to the ecologically important processes of denitrification, sulfate reduction, and acetogenesis, respectively.

Another way of life of chemotrophs in the absence of possible electron acceptors is fermentation, wherein the electrons taken from the reduced substrates are transferred to oxidised intermediates to generate reduced fermentation products (e.g., lactate, ethanol, hydrogen, butyric acid). Fermentation is possible, because the energy content of the substrates is higher than that of the products, which allows the organisms to synthesise ATP and drive their metabolism.

These processes are also important in biological responses to pollution; for example, sulfate-reducing bacteria are largely responsible for the production of the highly toxic forms of mercury (methyl- and dimethylmercury) in the environment. Non-respiratory anaerobes use fermentation to generate energy and reducing power, secreting metabolic by-products (such as ethanol in brewing) as waste. Facultative anaerobes can switch between fermentation and different terminal electron acceptors depending on the environmental conditions in which they find themselves.

Lithotrophic bacteria can use inorganic compounds as a source of energy. Common inorganic electron donors are hydrogen, carbon monoxide, ammonia (leading to nitrification), ferrous iron and other reduced metal ions, and several reduced sulfur compounds. In unusual circumstances, the gas methane can be used by methanotrophic bacteria as both a source of electrons and a substrate for carbon anabolism. In both aerobic phototrophy and chemolithotrophy, oxygen is used as a terminal electron acceptor, whereas under anaerobic conditions inorganic compounds are used instead. Most lithotrophic organisms are autotrophic, whereas organotrophic organisms are heterotrophic.

In addition to fixing carbon dioxide in photosynthesis, some bacteria also fix nitrogen gas (nitrogen fixation) using the enzyme nitrogenase. This environmentally important trait can be found in bacteria of nearly all the metabolic types listed above, but is not universal.

Regardless of the type of metabolic process they employ, the majority of bacteria are able to take in raw materials only in the form of relatively small molecules, which enter the cell by diffusion or through molecular channels in cell membranes. The Planctomycetes are the exception (as they are in possessing membranes around their nuclear materi-al). It has recently been shown that *Gemmata obscuriglobus* is able to take in large molecules via a process that in some ways resembles endocytosis, the process used by eukaryotic cells to engulf external items.

Growth and Reproduction

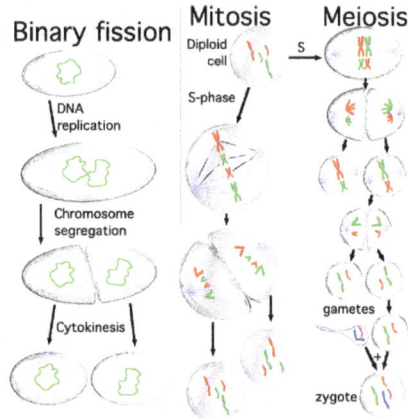

Many bacteria reproduce through binary fission, which is compared to mitosis and meiosis in this image.

Unlike in multicellular organisms, increases in cell size (cell growth) and reproduction by cell division are tightly linked in unicellular organisms. Bacteria grow to a fixed size and then reproduce through *binary fission*, a form of asexual reproduction. Under optimal conditions, bacteria can grow and divide extremely rapidly, and bacterial pop-ulations can double as quickly as every 9.8 minutes. In cell division, two identical clone daughter cells are produced. Some bacteria, while still reproducing asexually, form more complex reproductive structures that help disperse the newly formed daughter cells. Examples include fruiting body formation by *Myxobacteria* and aerial hyphae formation by *Streptomyces*, or budding. Budding involves a cell forming a protrusion that breaks away and produces a daughter cell.

A colony of *Escherichia coli*

In the laboratory, bacteria are usually grown using solid or liquid media. Solid *growth media*, such as agar plates, are used to isolate pure cultures of a bacterial strain.

However, liquid growth media are used when measurement of growth or large volumes of cells are required. Growth in stirred liquid media occurs as an even cell suspension, making the cultures easy to divide and transfer, although isolating single bacteria from liquid media is difficult. The use of selective media (media with specific nutrients added or deficient, or with antibiotics added) can help identify specific organisms.

Most laboratory techniques for growing bacteria use high levels of nutrients to produce large amounts of cells cheaply and quickly. However, in natural environments, nutrients are limited, meaning that bacteria cannot continue to reproduce indefinitely. This nutrient limitation has led the evolution of different growth strategies. Some organisms can grow extremely rapidly when nutrients become available, such as the formation of algal (and cyanobacterial) blooms that often occur in lakes during the summer. Other organisms have adaptations to harsh environments, such as the production of multiple antibiotics by *Streptomyces* that inhibit the growth of competing microorganisms. In nature, many organisms live in communities (e.g., biofilms) that may allow for increased supply of nutrients and protection from environmental stresses. These relationships can be essential for growth of a particular organism or group of organisms (syntrophy).

Bacterial growth follows four phases. When a population of bacteria first enter a high-nutrient environment that allows growth, the cells need to adapt to their new environment. The first phase of growth is the *lag phase*, a period of slow growth when the cells are adapting to the high-nutrient environment and preparing for fast growth. The lag phase has high biosynthesis rates, as proteins necessary for rapid growth are produced. The second phase of growth is the *log phase*, also known as the *logarithmic or exponential phase*. The log phase is marked by rapid exponential growth. The rate at which cells grow during this phase is known as the *growth rate* (k), and the time it takes the cells to double is known as the *generation time* (g). During log phase, nutrients are metabolised at maximum speed until one of the nutrients is depleted and starts limiting growth. The third phase of growth is the *stationary phase* and is caused by depleted nutrients. The cells reduce their metabolic activity and consume non-essential cellular proteins. The stationary phase is a transition from rapid growth to a stress response state and there is increased expression of genes involved in DNA repair, antioxidant metabolism and nutrient transport. The final phase is the *death phase* where the bacteria run out of nutrients and die.

Genomes

The genomes of thousands of bacterial species have been sequenced, with at least 9,000 sequences completed and more than 42,000 left as "permanent" drafts (as of Sep 2016).

Most bacteria have a single circular chromosome that can range in size from only 160,000 base pairs in the endosymbiotic bacteria *Candidatus Carsonella ruddii*, to

12,200,000 base pairs in the soil-dwelling bacteria *Sorangium cellulosum*. The genes in bacterial genomes are usually a single continuous stretch of DNA and although several different types of introns do exist in bacteria, these are much rarer than in eukaryotes. Some bacteria, including the Spirochaetes of the genus *Borrelia* are a notable exception to this arrangement. *Borrelia burgdorferi*, the cause of Lyme disease, contains a single linear chromosome and several linear and circular plasmids.

Plasmids are small extra-chromosomal DNAs that may contain genes for antibiotic resistance or virulence factors. Plasmids replicate independently of chromosomes, so it is possible that plasmids could be lost in bacterial cell division. Against this possibility is the fact that a single bacterium can contain hundreds of copies of a single plasmid.

Genetics

Bacteria, as asexual organisms, inherit identical copies of their parent's genes (i.e., they are clonal). However, all bacteria can evolve by selection on changes to their genetic material DNA caused by genetic recombination or mutations. Mutations come from errors made during the replication of DNA or from exposure to mutagens. Mutation rates vary widely among different species of bacteria and even among different clones of a single species of bacteria. Genetic changes in bacterial genomes come from either random mutation during replication or "stress-directed mutation", where genes involved in a particular growth-limiting process have an increased mutation rate.

DNA Transfer

Some bacteria also transfer genetic material between cells. This can occur in three main ways. First, bacteria can take up exogenous DNA from their environment, in a process called *transformation*. Genes can also be transferred by the process of *transduction*, when the integration of a bacteriophage introduces foreign DNA into the chromosome. The third method of gene transfer is *conjugation*, whereby DNA is transferred through direct cell contact.

Transduction of bacterial genes by bacteriophage appears to be a consequence of infrequent errors during intracellular assembly of virus particles, rather than a bacterial adaptation. Conjugation, in the much-studied E. coli system is determined by plasmid genes, and is an adaptation for transferring copies of the plasmid from one bacterial host to another. It is seldom that a conjugative plasmid integrates into the host bacterial chromosome, and subsequently transfers part of the host bacterial DNA to another bacterium. Plasmid-mediated transfer of host bacterial DNA also appears to be an accidental process rather than a bacterial adaptation.

Transformation, unlike transduction or conjugation, depends on numerous bacterial gene products that specifically interact to perform this complex process, and thus transformation is clearly a bacterial adaptation for DNA transfer. In order for a bacterium to bind, take up and recombine donor DNA into its own chromosome, it must first enter a special physiological state termed competence. In *Bacillus subtilis*, about 40 genes are required for the development of competence. The length of DNA transferred during *B. subtilis* transformation can be between a third of a chromosome up to the whole chromosome. Transformation appears to be common among bacterial species, and thus far at least 60 species are known to have the natural ability to become competent for transformation. The development of competence in nature is usually associated with stressful environmental conditions, and seems to be an adaptation for facilitating repair of DNA damage in recipient cells.

In ordinary circumstances, transduction, conjugation, and transformation involve transfer of DNA between individual bacteria of the same species, but occasionally transfer may occur between individuals of different bacterial species and this may have significant consequences, such as the transfer of antibiotic resistance. In such cases, gene acquisition from other bacteria or the environment is called *horizontal gene transfer* and may be common under natural conditions. Gene transfer is particularly important in antibiotic resistance as it allows the rapid transfer of resistance genes between different pathogens.

Bacteriophages

Bacteriophages are viruses that infect bacteria. Many types of bacteriophage exist, some simply infect and lyse their host bacteria, while others insert into the bacterial chromosome. A bacteriophage can contain genes that contribute to its host's phenotype: for example, in the evolution of *Escherichia coli* O157:H7 and *Clostridium botulinum*, the toxin genes in an integrated phage converted a harmless ancestral bacterium into a lethal pathogen. Bacteria resist phage infection through restriction modification systems that degrade foreign DNA, and a system that uses CRISPR sequences to retain fragments of the genomes of phage that the bacteria have come into contact with in the past, which allows them to block virus replication through a form of RNA interference. This CRISPR system provides bacteria with acquired immunity to infection.

Behaviour

Secretion

Bacteria frequently secrete chemicals into their environment in order to modify it favourably. The secretions are often proteins and may act as enzymes that digest some form of food in the environment.

Bioluminescence

A few bacteria have chemical systems that generate light. This bioluminescence often occurs in bacteria that live in association with fish, and the light probably serves to attract fish or other large animals.

Multicellularity

Bacteria often function as multicellular aggregates known as biofilms, exchanging a variety of molecular signals for inter-cell communication, and engaging in coordinated multicellular behaviour.

The communal benefits of multicellular cooperation include a cellular division of labour, accessing resources that cannot effectively be used by single cells, collectively defending against antagonists, and optimising population survival by differentiating into distinct cell types. For example, bacteria in biofilms can have more than 500 times increased resistance to antibacterial agents than individual "planktonic" bacteria of the same species.

One type of inter-cellular communication by a molecular signal is called quorum sensing, which serves the purpose of determining whether there is a local population density that is sufficiently high that it is productive to invest in processes that are only successful if large numbers of similar organisms behave similarly, as in excreting digestive enzymes or emitting light.

Quorum sensing allows bacteria to coordinate gene expression, and enables them to produce, release and detect autoinducers or pheromones which accumulate with the growth in cell population.

Movement

Many bacteria can move using a variety of mechanisms: flagella are used for swimming through fluids; bacterial gliding and twitching motility move bacteria across surfaces; and changes of buoyancy allow vertical motion.

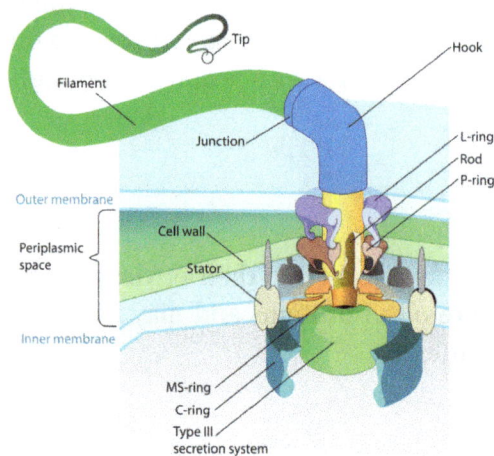

Flagellum of gram-negative bacteria. The base drives the rotation of the hook and filament.

Swimming bacteria frequently move near 10 body lengths per second and a few as fast as 100. This makes them at least as fast as fish, on a relative scale.

In bacterial gliding and twitching motility, bacteria use their *type IV pili* as a grappling hook, repeatedly extending it, anchoring it and then retracting it with remarkable force (>80 pN).

"Our observations redefine twitching motility as a rapid, highly organized mechanism of bacterial translocation by which Pseudomonas aeruginosa can disperse itself over large areas to colonize new territories. It is also now clear, both morphologically and genetically, that twitching motility and social gliding motility, such as occurs in Myxococcus xanthus, are essentially the same process."

— *"A re-examination of twitching motility in Pseudomonas aeruginosa" – Semmler, Whitchurch & Mattick (1999)*

Flagella are semi-rigid cylindrical structures that are rotated and function much like the propeller on a ship. Objects as small as bacteria operate a low Reynolds number and cylindrical forms are more efficient than the flat, paddle-like, forms appropriate at human-size scale.

Bacterial species differ in the number and arrangement of flagella on their surface; some have a single flagellum (*monotrichous*), a flagellum at each end (*amphitrichous*), clusters of flagella at the poles of the cell (*lophotrichous*), while others have flagella distributed over the entire surface of the cell (*peritrichous*). The bacterial flagella is the best-understood motility structure in any organism and is made of about 20 proteins, with approximately another 30 proteins required for its regulation and assembly. The flagellum is a rotating structure driven by a reversible motor at the base that uses the electrochemical gradient across the membrane for power. This motor drives the motion of the filament, which acts as a propeller.

Many bacteria (such as *E. coli*) have two distinct modes of movement: forward movement (swimming) and tumbling. The tumbling allows them to reorient and makes their movement a three-dimensional random walk. The flagella of a unique group of bacteria, the spirochaetes, are found between two membranes in the periplasmic space. They have a distinctive helical body that twists about as it moves.

Motile bacteria are attracted or repelled by certain stimuli in behaviours called taxes: these include chemotaxis, phototaxis, energy taxis, and magnetotaxis. In one peculiar group, the myxobacteria, individual bacteria move together to form waves of cells that then differentiate to form fruiting bodies containing spores. The myxobacteria move only when on solid surfaces, unlike *E. coli*, which is motile in liquid or solid media.

Several *Listeria* and *Shigella* species move inside host cells by usurping the cytoskeleton, which is normally used to move organelles inside the cell. By promoting actin polymerisation at one pole of their cells, they can form a kind of tail that pushes them through the host cell's cytoplasm.

Classification and Identification

Classification seeks to describe the diversity of bacterial species by naming and grouping organisms based on similarities. Bacteria can be classified on the basis of cell structure, cellular metabolism or on differences in cell components, such as DNA, fatty acids,
pigments, antigens and quinones. While these schemes allowed the identification and
classification of bacterial strains, it was unclear whether these differences represented
variation between distinct species or between strains of the same species. This uncertainty was due to the lack of distinctive structures in most bacteria, as well as lateral
gene transfer between unrelated species. Due to lateral gene transfer, some closely related bacteria can have very different morphologies and metabolisms. To overcome this
uncertainty, modern bacterial classification emphasises molecular systematics, using
genetic techniques such as guanine cytosine ratio determination, genome-genome hybridisation, as well as sequencing genes that have not undergone extensive lateral gene
transfer, such as the rRNA gene. Classification of bacteria is determined by publication
in the International Journal of Systematic Bacteriology, and Bergey's Manual of Systematic Bacteriology. The International Committee on Systematic Bacteriology (ICSB)
maintains international rules for the naming of bacteria and taxonomic categories and
for the ranking of them in the International Code of Nomenclature of Bacteria.

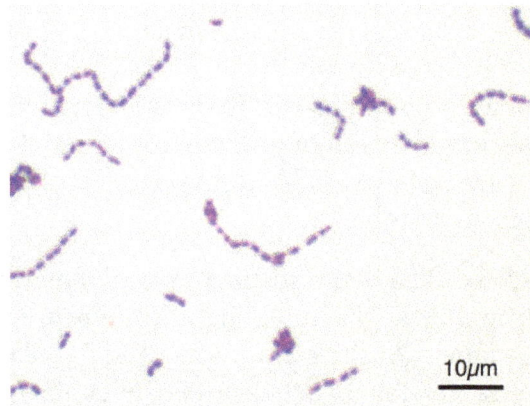

Streptococcus mutans visualised with a Gram stain

The term "bacteria" was traditionally applied to all microscopic, single-cell prokaryotes. However, molecular systematics showed prokaryotic life to consist of two separate
domains, originally called *Eubacteria* and *Archaebacteria*, but now called *Bacteria*
and *Archaea* that evolved independently from an ancient common ancestor. The archaea and eukaryotes are more closely related to each other than either is to the bacteria. These two domains, along with Eukarya, are the basis of the three-domain system, which is currently the most widely used classification system in microbiolology.
However, due to the relatively recent introduction of molecular systematics and a rapid
increase in the number of genome sequences that are available, bacterial classification
remains a changing and expanding field. For example, a few biologists argue that the
Archaea and Eukaryotes evolved from gram-positive bacteria.

The identification of bacteria in the laboratory is particularly relevant in medicine, where the correct treatment is determined by the bacterial species causing an infection. Consequently, the need to identify human pathogens was a major impetus for the development of techniques to identify bacteria.

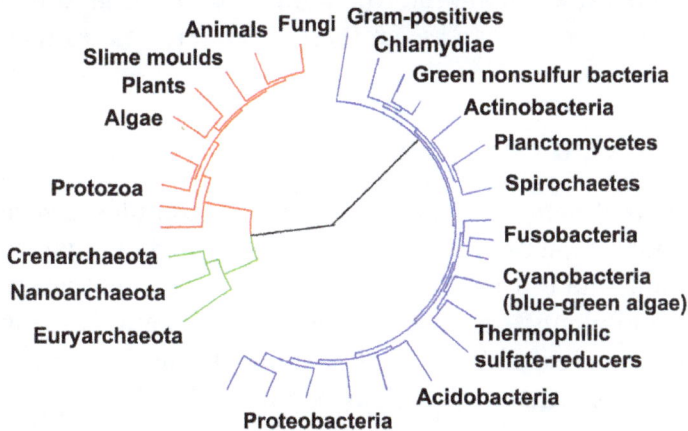

Phylogenetic tree showing the diversity of bacteria, compared to other organisms. Eukaryotes are coloured red, archaea green and bacteria blue.

The *Gram stain*, developed in 1884 by Hans Christian Gram, characterises bacteria based on the structural characteristics of their cell walls. The thick layers of peptidoglycan in the "gram-positive" cell wall stain purple, while the thin "gram-negative" cell wall appears pink. By combining morphology and Gram-staining, most bacteria can be classified as belonging to one of four groups (gram-positive cocci, gram-positive bacilli, gram-negative cocci and gram-negative bacilli). Some organisms are best identified by stains other than the Gram stain, particularly mycobacteria or *Nocardia*, which show acid-fastness on Ziehl–Neelsen or similar stains. Other organisms may need to be identified by their growth in special media, or by other techniques, such as serology.

Culture techniques are designed to promote the growth and identify particular bacteria, while restricting the growth of the other bacteria in the sample. Often these techniques are designed for specific specimens; for example, a sputum sample will be treated to identify organisms that cause pneumonia, while stool specimens are cultured on selective media to identify organisms that cause diarrhoea, while preventing growth of non-pathogenic bacteria. Specimens that are normally sterile, such as blood, urine or spinal fluid, are cultured under conditions designed to grow all possible organisms. Once a pathogenic organism has been isolated, it can be further characterised by its morphology, growth patterns (such as aerobic or anaerobic growth), patterns of hemolysis, and staining.

As with bacterial classification, identification of bacteria is increasingly using molecular methods. Diagnostics using DNA-based tools, such as polymerase chain reaction, are increasingly popular due to their specificity and speed, compared to culture-based methods. These methods also allow the detection and identification of "viable but nonculturable" cells that are metabolically active but non-dividing. However, even using these improved methods, the total number of bacterial species is not known and cannot even be estimated with any certainty. Following present classification, there are a little less than 9,300 known species of prokaryotes, which includes bacteria and archaea; but attempts to estimate the true number of bacterial diversity have ranged from 10^7 to 10^9 total species – and even these diverse estimates may be off by many orders of magnitude.

Interactions with Other Organisms

Despite their apparent simplicity, bacteria can form complex associations with other organisms. These symbiotic associations can be divided into parasitism, mutualism and commensalism. Due to their small size, commensal bacteria are ubiquitous and grow on animals and plants exactly as they will grow on any other surface. However, their growth can be increased by warmth and sweat, and large populations of these organisms in humans are the cause of body odour.

Predators

Some species of bacteria kill and then consume other microorganisms, these species are called *predatory bacteria*. These include organisms such as *Myxococcus xanthus*, which forms swarms of cells that kill and digest any bacteria they encounter. Other bacterial predators either attach to their prey in order to digest them and absorb nutrients, such as *Vampirovibrio chlorellavorus*, or invade another cell and multiply inside the cytosol, such as *Daptobacter*. These predatory bacteria are thought to have evolved from saprophages that consumed dead microorganisms, through adaptations that allowed them to entrap and kill other organisms.

Mutualists

Certain bacteria form close spatial associations that are essential for their survival. One such mutualistic association, called interspecies hydrogen transfer, occurs between clusters of anaerobic bacteria that consume organic acids, such as butyric acid or propionic acid, and produce hydrogen, and methanogenic Archaea that consume hydrogen. The bacteria in this association are unable to consume the organic acids as this reaction produces hydrogen that accumulates in their surroundings. Only the intimate association with the hydrogen-consuming Archaea keeps the hydrogen concentration low enough to allow the bacteria to grow.

In soil, microorganisms that reside in the rhizosphere (a zone that includes the root surface and the soil that adheres to the root after gentle shaking) carry out nitrogen fixation, converting nitrogen gas to nitrogenous compounds. This serves to provide an easily absorbable form of nitrogen for many plants, which cannot fix nitrogen themselves. Many other bacteria are found as symbionts in humans and other organisms.

For example, the presence of over 1,000 bacterial species in the normal human gut flora of the intestines can contribute to gut immunity, synthesise vitamins, such as folic acid, vitamin K and biotin, convert sugars to lactic acid, as well as fermenting complex undigestible carbohydrates. The presence of this gut flora also inhibits the growth of potentially pathogenic bacteria (usually through competitive exclusion) and these beneficial bacteria are consequently sold as probiotic dietary supplements.

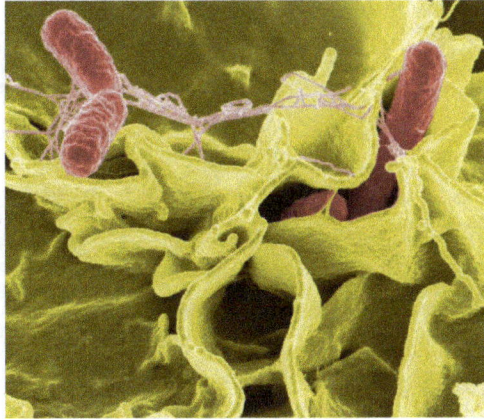

Colour-enhanced scanning electron micrograph showing *Salmonella typhimurium* (red) invading cultured human cells

Pathogens

If bacteria form a parasitic association with other organisms, they are classed as pathogens. Pathogenic bacteria are a major cause of human death and disease and cause infections such as tetanus, typhoid fever, diphtheria, syphilis, cholera, foodborne illness, leprosy and tuberculosis. A pathogenic cause for a known medical disease may only be discovered many years after, as was the case with *Helicobacter pylori* and peptic ulcer disease. Bacterial diseases are also important in agriculture, with bacteria causing leaf spot, fire blight and wilts in plants, as well as Johne's disease, mastitis, salmonella and anthrax in farm animals.

Each species of pathogen has a characteristic spectrum of interactions with its human hosts. Some organisms, such as *Staphylococcus* or *Streptococcus*, can cause skin infections, pneumonia, meningitis and even overwhelming sepsis, a systemic inflammatory response producing shock, massive vasodilation and death. Yet these organisms are also part of the normal human flora and usually exist on the skin or in the nose without causing any disease at all. Other organisms invariably cause disease in humans, such as the Rickettsia, which are obligate intracellular parasites able to grow and reproduce only within the cells of other organisms. One species of Rickettsia causes typhus, while another causes Rocky Mountain spotted fever. *Chlamydia*, another phylum of obligate intracellular parasites, contains species that can cause pneumonia, or urinary tract infection and may be involved in coronary heart disease. Finally, some species, such as

Pseudomonas aeruginosa, *Burkholderia cenocepacia*, and *Mycobacterium avium*, are opportunistic pathogens and cause disease mainly in people suffering from immuno-suppression or cystic fibrosis.

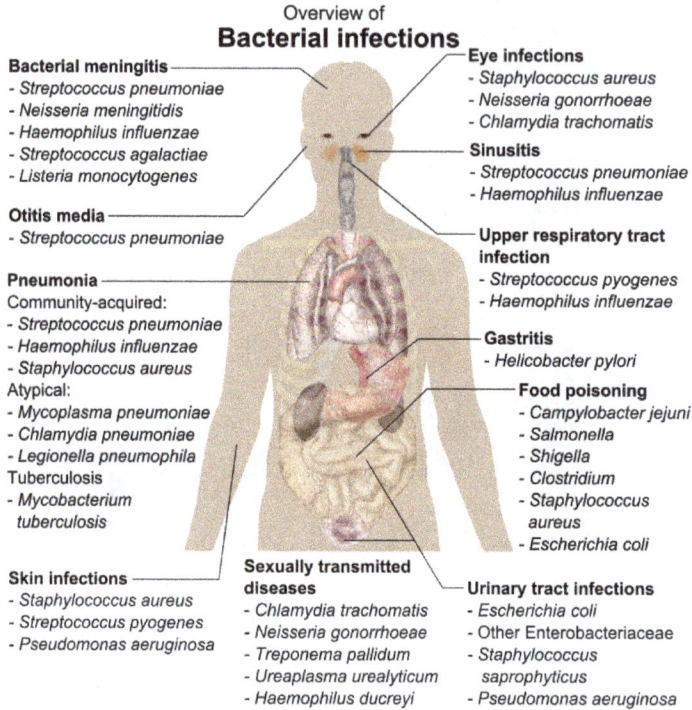

Overview of bacterial infections and main species involved.

Bacterial infections may be treated with antibiotics, which are classified as bacterio-cidal if they kill bacteria, or bacteriostatic if they just prevent bacterial growth. There are many types of antibiotics and each class inhibits a process that is different in the pathogen from that found in the host. An example of how antibiotics produce selective toxicity are chloramphenicol and puromycin, which inhibit the bacterial ribosome, but not the structurally different eukaryotic ribosome. Antibiotics are used both in treating human disease and in intensive farming to promote animal growth, where they may be contributing to the rapid development of antibiotic resistance in bacterial populations. Infections can be prevented by antiseptic measures such as sterilising the skin prior to piercing it with the needle of a syringe, and by proper care of indwelling catheters. Surgical and dental instruments are also sterilised to prevent contamination by bacteria. Disinfectants such as bleach are used to kill bacteria or other pathogens on surfaces to prevent contamination and further reduce the risk of infection.

Significance in Technology and Industry

Bacteria, often lactic acid bacteria, such as *Lactobacillus* and *Lactococcus*, in combination with yeasts and moulds, have been used for thousands of years in the preparation of fermented foods, such as cheese, pickles, soy sauce, sauerkraut, vinegar, wine and yogurt.

The ability of bacteria to degrade a variety of organic compounds is remarkable and has been used in waste processing and bioremediation. Bacteria capable of digesting the hydrocarbons in petroleum are often used to clean up oil spills. Fertiliser was added to some of the beaches in Prince William Sound in an attempt to promote the growth of these naturally occurring bacteria after the 1989 *Exxon Valdez* oil spill. These efforts were effective on beaches that were not too thickly covered in oil. Bacteria are also used for the bioremediation of industrial toxic wastes. In the chemical industry, bacteria are most important in the production of enantiomerically pure chemicals for use as pharmaceuticals or agrichemicals.

Bacteria can also be used in the place of pesticides in the biological pest control. This commonly involves *Bacillus thuringiensis* (also called BT), a gram-positive, soil dwelling bacterium. Subspecies of this bacteria are used as a Lepidopteran-specific insecticides under trade names such as Dipel and Thuricide. Because of their specificity, these pesticides are regarded as environmentally friendly, with little or no effect on humans, wildlife, pollinators and most other beneficial insects.

Because of their ability to quickly grow and the relative ease with which they can be manipulated, bacteria are the workhorses for the fields of molecular biology, genetics and biochemistry. By making mutations in bacterial DNA and examining the resulting phenotypes, scientists can determine the function of genes, enzymes and metabolic pathways in bacteria, then apply this knowledge to more complex organisms. This aim of understanding the biochemistry of a cell reaches its most complex expression in the synthesis of huge amounts of enzyme kinetic and gene expression data into mathematical models of entire organisms. This is achievable in some well-studied bacteria, with models of *Escherichia coli* metabolism now being produced and tested. This understanding of bacterial metabolism and genetics allows the use of biotechnology to bioengineer bacteria for the production of therapeutic proteins, such as insulin, growth factors, or antibodies.

Because of their importance for research in general, samples of bacterial strains are isolated and preserved in Biological Resource Centers. This ensures the availability of the strain to scientists worldwide.

History of Bacteriology

Bacteria were first observed by the Dutch microscopist Antonie van Leeuwenhoek in 1676, using a single-lens microscope of his own design. He then published his observations in a series of letters to the Royal Society of London. Bacteria were Leeuwenhoek's most remarkable microscopic discovery. They were just at the limit of what his simple lenses could make out and, in one of the most striking hiatuses in the history of science, no one else would see them again for over a century. Only then were his by-then-largely-forgotten observations of bacteria — as opposed to his famous "animalcules" (spermatozoa) — taken seriously.

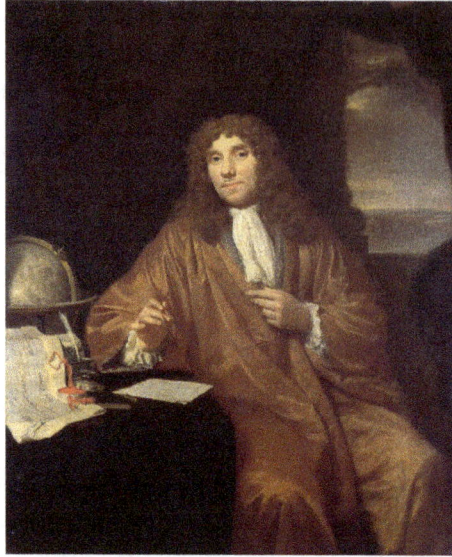

Antonie van Leeuwenhoek, the first microbiologist and the first person
to observe bacteria using a microscope.

Christian Gottfried Ehrenberg introduced the word "bacterium" in 1828. In fact, his *Bacterium* was a genus that contained non-spore-forming rod-shaped bacteria, as opposed to *Bacillus*, a genus of spore-forming rod-shaped bacteria defined by Ehrenberg in 1835.

Louis Pasteur demonstrated in 1859 that the growth of microorganisms causes the fermentation process, and that this growth is not due to spontaneous generation. (Yeasts and moulds, commonly associated with fermentation, are not bacteria, but rather fungi.) Along with his contemporary Robert Koch, Pasteur was an early advocate of the germ theory of disease.

Robert Koch, a pioneer in medical microbiology, worked on cholera, anthrax and tuberculosis. In his research into tuberculosis Koch finally proved the germ theory, for which he received a Nobel Prize in 1905. In *Koch's postulates*, he set out criteria to test if an organism is the cause of a disease, and these postulates are still used today.

Though it was known in the nineteenth century that bacteria are the cause of many diseases, no effective antibacterial treatments were available. In 1910, Paul Ehrlich developed the first antibiotic, by changing dyes that selectively stained *Treponema pallidum* — the spirochaete that causes syphilis — into compounds that selectively killed the pathogen. Ehrlich had been awarded a 1908 Nobel Prize for his work on immunology, and pioneered the use of stains to detect and identify bacteria, with his work being the basis of the Gram stain and the Ziehl–Neelsen stain.

A major step forward in the study of bacteria came in 1977 when Carl Woese recognised that archaea have a separate line of evolutionary descent from bacteria. This new phylogenetic taxonomy depended on the sequencing of 16S ribosomal RNA, and divided prokaryotes into two evolutionary domains, as part of the three-domain system.

References

- Jeanne Stove Poindexter (30 November 1986). Methods and special applications in bacterial ecology. Springer. p. 87. ISBN 978-0-306-42346-8. Retrieved 18 June 2011.

- Dusenbery, David B. (2009). Living at Micro Scale, pp. 20–25. Harvard University Press, Cambridge, Mass. ISBN 978-0-674-03116-6.

- Hecker M, Völker U (2001). "General stress response of Bacillus subtilis and other bacteria". Adv Microb Physiol. Advances in Microbial Physiology. 44: 35–91. doi:10.1016/S0065-2911(01)44011-2. ISBN 978-0-12-027744-5.

- Fisher B, Harvey RP, Champe PC (2007). Lippincott's Illustrated Reviews: Microbiology (Lippincott's Illustrated Reviews Series). Hagerstwon, MD: Lippincott Williams & Wilkins. pp. Chapter 33, pages 367–392. ISBN 0-7817-8215-5.

- The University of Waikato (March 25, 2014). "Bacterial DNA – the role of plasmids". Themes — Bacteria in biotech. Biotechnology Learning Hub. Retrieved 2014-09-03.

- Euzéby JP (8 December 2011). "Number of published names". List of Prokaryotic names with Standing in Nomenclature. Archived from the original on 19 January 2012. Retrieved 10 December 2011.

- Wassenaar, T. M. "Bacteriology: the study of bacteria". www.mmgc.eu. Archived from the original on 24 July 2011. Retrieved 18 June 2011.

- Ward J. MacNeal; Herbert Upham Williams (1914). Pathogenic micro-organisms; a text-book of microbiology for physicians and students of medicine. P. Blakiston's sons & co. pp. 1–. Retrieved 18 June 2011.

Bacterial Taxonomy: An Integrated Study

Bacterial taxonomy is the classification of bacteria on the basis of their rank in the taxonomic hierarchy. Some of the examples of these ranks are genus, family, kingdom etc. Topics such as monera and bacterial phyla have also been elucidated in the following section. The topics discussed in the chapter are of great importance to broaden the existing knowledge on bacteriology.

Bacterial Taxonomy

Bacterial taxonomy is the taxonomy, i.e. the rank-based classification, of bacteria.

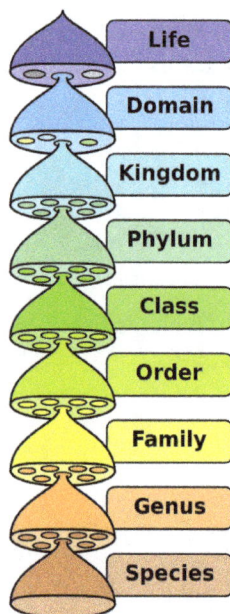

The hierarchy of biological classification's eight major taxonomic ranks.
Intermediate minor rankings are not shown.

In the scientific classification established by Carl von Linné, each species has to be assigned to a genus (binary nomenclature), which in turn is a lower level of a hierarchy of ranks (family, suborder, order, subclass, class, division/phyla, kingdom and domain). In the currently accepted classification of life, there are three domains (Eukaryotes, Bacteria and Archaea), which, in terms of taxonomy, despite following the same principles have several different conventions between them and between their subdivisions as are studied by different disciplines (botany, zoology, mycology and microbiology), for example in zoology there are type specimens, whereas in microbiology there are type strains.

Diversity

Prokaryotes share many common features, such as lack of nuclear membrane, unicellularity, division by binary-fission and generally small size. The various species differ amongst each other based on several characteristics, allowing their identification and classification. Examples include:

- Phylogeny: All bacteria stem from a common ancestor and diversified since, consequently possess different levels of evolutionary relatedness

- Metabolism: Different bacteria may have different metabolic abilities

- Environment: Different bacteria thrive in different environments, such as high/low temperature and salt

- Morphology: There are many structural differences between bacteria, such as cell shape, Gram stain (number of lipid bilayers) or bilayer composition

- Pathogenicity: Some bacteria are pathogenic to plants or animals

Classification History

Bacteria were first observed by Antonie van Leeuwenhoek in 1676, using a single-lens microscope of his own design. He called them "animalcules" and published his observations in a series of letters to the Royal Society. O. F. Müller (1773, 1786) described eight species of the genus *Vibrio* (in Infusoria), three of which were spirilliforms. The term *Bacterium* (a genus) was introduced much later, by Christian Gottfried Ehrenberg in 1838.

Classical Classification

Placement

Bacteria were first classified as plants constituting the class *Schizomycetes*, which along with the *Schizophyceae* (blue green algae/*Cyanobacteria*) formed the phylum *Schizophyta*.

Haeckel in 1866 placed the group in the phylum *Moneres* in the kingdom *Protista* and defines them as completely structureless and homogeneous organisms, consisting only of a piece of plasma. He subdivided the phylum into two groups:

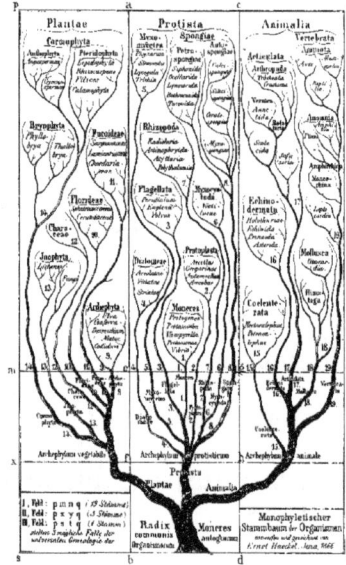

Tree of Life in Generelle Morphologie der Organismen (1866)

- die Gymnomoneren (no envelope)

  - *Protogenes*—such as *Protogenes primordialis*, now classed as a eukaryote and not a bacterium

  - *Protamaeba*—now classed as a eukaryote and not a bacterium

  - *Vibrio*—a genus of comma shaped bacteria first described in 1854)

  - *Bacterium*—a genus of rod shaped bacteria first described in 1828, that later gave its name to the members of the Monera, formerly referred to as "a moneron" (plural "monera") in English and "eine Moneren"(fem. plural "Moneres") in German

  - *Bacillus*—a genus of spore-forming rod shaped bacteria first described in 1835

  - *Spirochaeta*—thin spiral shaped bacteria first described in 1835

  - *Spirillum*—spiral shaped bacteria first described in 1832

  - etc.

- die Lepomoneren (with envelope)

  - *Protomonas*—now classed as a eukaryote and not a bacterium. The name was reused in 1984 for an unrelated genus of Bacteria

  - *Vampyrella*—now classed as a eukaryote and not a bacterium

The group was later reclassified as the *Prokaryotes* by Chatton.

The classification of *Cyanobacteria* (colloquially "blue green algae") has been fought between being algae or bacteria (for example, Haeckel classified *Nostoc* in the phylum Archephyta of Algae).

in 1905 Erwin F. Smith accepted 33 valid different names of bacterial genera and over 150 invalid names, and in 1913 Vuillemin in a study concluded that all species of the *Bacteria* should fall into the genera *Planococcus, Streptococcus, Klebsiella, Merista, Planomerista, Neisseria, Sarcina, Planosarcina, Metabacterium, Clostridium, Serratia, Bacterium* and *Spirillum*.

Ferdinand Cohn recognized 4 tribes: Spherobacteria, Microbacteria, Desmobacteria, and Spirobacteria. Stanier and van Neil recognized the Kingdom Monera with 2 phyla, Myxophyta and Schizomycetae, the latter comprising classes Eubacteriae (3 orders), Myxobacteriae (1 order), and Spirochetae (1 order). Bisset distinguished 1 class and 4 orders: Eubacteriales, Actinomycetales, Streptomycetales, and Flexibacteriales. Migula, which was the most widely accepted system of its time and included all then-known species but was based only on morphology, contained the 3 basic groups, Coccaceae, Bacillaceae, and Spirillaceae but also Trichobacterinae for filamentous bacteria; Orla-Jensen established 2 orders: Cephalotrichinae (7 families) and Peritrichinae (presumably with only 1 family). Bergey et al presented a classification which generally followed the 1920 Final Report of the SAB(Society of American Bacteriologists) Committee (Winslow et al), which divided Class Schizomycetes into 4 orders: Myxobacteriales, Thiobacteriales, Chlamydobacteriales, and Eubacteriales, with a 5th group being 4 genera considered intermediate between bacteria and protozoans: *Spirocheta, Cristospira, Saprospira,* and *Treponema*.

However, different authors often reclassified the genera due to the lack of visible traits to go by, resulting in a poor state which was summarised in 1915 by Robert Earle Buchanan By then, the whole group received different ranks and names by different authors namely

- *Schizomycetes* (Naegeli 1857)

- *Bacteriaceae* (Cohn 1872,)

- *Bacteria* (Cohn 1872b,)

- *Schizomycetaceae* (DeToni and Trevisan 1889,)

Furthermore the families into which the class was subdivided, changed from author to author and for some such as Zipf the names where in German and not in Latin The first edition of the Bacteriological Code in 1947 sorted several problems out.

A.R. Prévot's system) had 4 subphyla and 8 classes as follows:

Eubacteriales (classes Asporulales and Sporulales) Mycobacteriales (classes Actinomycetales, Myxobacteriales, and Azotobacteriales) Algobacteriales (classes Siderobacteriales and Thiobacteriales) Protozoobacteriales (class Spirochetales)

Linnaeus 1735	Haeckel 1866	Chatton 1925	Copeland 1938	Whittaker 1969	Woese et al. 1990	Cavalier-Smith 1998
2 kingdoms	3 kingdoms	2 empires	4 kingdoms	5 kingdoms	3 domains	6 kingdoms
(not treated)	Protista	Prokaryota	Monera	Monera	Bacteria	Bacteria
					Archaea	Bacteria
		Eukaryota	Protoctista	Protista	Eucarya	Protozoa
						Chromista
Vegetabilia	Plantae		Plantae	Plantae		Plantae
				Fungi		Fungi
Animalia	Animalia		Animalia	Animalia		Animalia

Subdivisions Based on Gram Staining

Despite there being little agreement on the major subgroups of the *Bacteria*, Gram staining results were most commonly used as a classification tool. Consequently, until the advent of molecular phylogeny, the Kingdom *Prokaryotae* was divided into four divisions, A classification scheme still formally followed by Bergey's manual of systematic bacteriology for tome order

- *Gracilicutes* (gram-negative)

    o *Photobacteria* (photosynthetic): class *Oxyphotobacteriae* (water as electron donor, includes the order *Cyanobacteriales*=blue-green algae, now phylum *Cyanobacteria*) and class *Anoxyphotobacteriae* (anaerobic phototrophs, orders: *Rhodospirillales* and *Chlorobiales*

    o *Scotobacteria* (non-photosynthetic, now the *Proteobacteria* and other gram-negative nonphotosynthetic phyla)

- *Firmacutes* [sic] (gram-positive, subsequently corrected to *Firmicutes*)

    o several orders such as *Bacillales* and *Actinomycetales* (now in the phylum *Actinobacteria*)

- *Mollicutes* (gram variable, e.g. *Mycoplasma*)

- *Mendocutes* (uneven gram stain, "methanogenic bacteria", now known as the *Archaea*)

Molecular Era

"Archaic Bacteria" and Woese's Reclassification

Woese argued that the bacteria, archaea, and eukaryotes represent separate lines of descent that diverged early on from an ancestral colony of organisms. However, a few biologists argue that the Archaea and Eukaryota arose from a group of bacteria. In any case, it is thought that viruses and archaea began relationships approximately two billion years ago, and that co-evolution may have been occurring between members of these groups. It is possible that the last common ancestor of the bacteria and archaea was a thermophile, which raises the possibility that lower temperatures are "extreme environments" in archaeal terms, and organisms that live in cooler environments appeared only later. Since the Archaea and Bacteria are no more related to each other than they are to eukaryotes, the term *prokaryote's* only surviving meaning is "not a eukaryote", limiting its value.

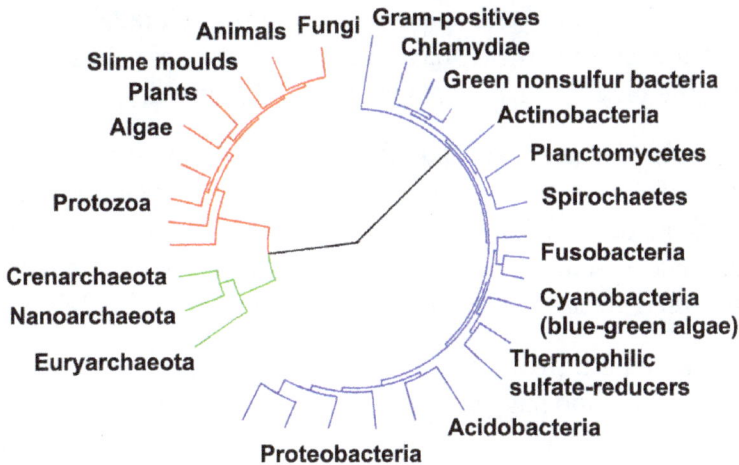

Phylogenetic tree showing the relationship between the archaea and other forms of life. Eukaryotes are colored red, archaea green and bacteria blue. Adapted from Ciccarelli *et al.*

With improved methodologies it became clear that the methanogenic bacteria were profoundly different and were (erroneously) believed to be relics of ancient bacteria thus Carl Woese, regarded as the forerunner of the molecular phylogeny revolution, identified three primary lines of descent: the *Archaebacteria*, the *Eubacteria*, and the *Urkaryotes*, the latter now represented by the nucleocytoplasmic component of the *Eukaryotes*. These lineages were formalised into the rank Domain (*regio* in Latin) which divided Life into 3 domains: the *Eukaryota*, the *Archaea* and the *Bacteria*. This scheme is still followed today.

Subdivisions

In 1987 Carl Woese divided the *Eubacteria* into 11 divisions based on 16S ribosomal RNA (SSU) sequences, which with several additions are still used today.

Opposition

While the three domain system is widely accepted, some authors have opposed it for various reasons.

One prominent scientist who opposes the three domain system is Thomas Cavalier-Smith, who proposed that the *Archaea* and the *Eukaryotes* (the *Neomura*) stem from Gram positive bacteria (*Posibacteria*), which in turn derive from gram negative bacteria (*Negibacteria*) based on several logical arguments, which are highly controversial and generally disregarded by the molecular biology community (*c.f.* reviewers' comments on, *e.g.* Eric Bapteste is "agnostic" regarding the conclusions) and are often not mentioned in reviews (*e.g.*) due to the subjective nature of the assumptions made.

However, despite there being a wealth of statistically supported studies towards the rooting of the tree of life between the *Bacteria* and the *Neomura* by means of a variety of methods, including some that are impervious to accelerated evolution—which is claimed by Cavalier-Smith to be the source of the supposed fallacy in molecular methods—there are a few studies which have drawn different conclusions, some of which place the root in the phylum *Firmicutes* with nested archaea.

Radhey Gupta's molecular taxonomy, based on conserved signature sequences of proteins, includes a monophyletic Gram negative clade, a monophyletic Gram positive clade, and a polyphyletic Archeota derived from Gram positives. Hori and Osawa's molecular analysis indicated a link between Metabacteria (=Archeota) and eukaryotes. The only cladistic analyses for bacteria based on classical evidence largely corroborate Gupta's results.

James Lake presented a 2 primary kingdom arrangement (Parkaryotae + eukaryotes and eocytes + Karyotae) and suggested a 5 primary kingdom scheme (Eukaryota, Eocyta, Methanobacteria, Halobacteria, and Eubacteria) based on ribosomal structure and a 4 primary kingdom scheme (Eukaryota, Eocyta, Methanobacteria, and Photocyta), bacteria being classified according to 3 major biochemical innovations: photosynthesis (Photocyta), methanogenesis (Methanobacteria), and sulfur respiration (Eocyta). He has also discovered evidence that Gram-negative bacteria arose from a symbiosis between 2 Gram-positive bacteria.

Authorities

Classification is the grouping of organisms into progressively more inclusive groups based on phylogeny and phenotype, while nomenclature is the application of formal rules for naming organisms.

Nomenclature Authority

Despite there being no official and complete classification of prokaryotes, the names (nomenclature) given to prokaryotes are regulated by the International Code of

Nomenclature of Bacteria (Bacteriological Code), a book which contains general considerations, principles, rules, and various notes, and advises in a similar fashion to the nomenclature codes of other groups.

Classification Authorities

The taxa which have been correctly described are reviewed in *Bergey's manual of Systematic Bacteriology*, which aims to aid in the identification of species and is considered the highest authority.

LPSN is an online database which currently contains over two thousand accepted names with their references, etymologies and various notes.

Description of New Species

The *International Journal of Systematic Bacteriology/International Journal of Systematic and Evolutionary Microbiology* (IJSB/IJSEM) is a peer reviewed journal which acts as the official international forum for the publication of new prokaryotic taxa. If a species is published in a different peer review journal, the author can submit a request to IJSEM with the appropriate description, which if correct, the new species will be featured in the Validation List of IJSEM.

Distribution

Microbial culture collections are depositories of strains which aim to safeguard them and to distribute them. The main ones being:

Collection Acronym	Name	Location
ATCC	American Type Culture Collection	Manassas, Virginia
NCTC	National Collection of Type Cultures	Public Health England, United Kingdom
BCCM	Belgium Coordinated Collection of Microorganisms	Ghent, Belgium
CIP	Collection d'Institut Pasteur	Paris, France
DSMZ	Deutsche Sammlung von Mikroorganismen und Zellkulturen	Braunschweig, Germany
JCM	Japan Collection of Microorganisms	Saitama, Japan
NCCB	Netherlands Culture Collection of Bacteria	Utrecht, Netherlands
NCIMB	National Collection of industrial, Marine and food bacteria	Aberdeen, Scotland
ICMP	International Collection of Microorganisms from Plants	Auckland, New Zealand

Analyses

Bacteria were at first classified based solely on their shape (vibrio, bacillus, coccus etc.), presence of endospores, gram stain, aerobic conditions and motility. This system changed with the study of metabolic phenotypes, where metabolic characteristics were used. Recently, with the advent of molecular phylogeny, several genes are used to identify species, the most important of which is the 16S rRNA gene, followed by 23S, ITS region, gyrB and others to confirm a better resolution. The quickest way to identify to match an isolated strain to a species or genus today is done by amplifying it's 16S gene with universal primers and sequence the 1.4kb amplicon and submit it to a specialised web-based identification database, namely either Ribosomal Database Project, which align the sequence to other 16S sequences using infernal, a secondary structure bases global alignment, or ARB SILVA, which aligns sequences via SINA (SILVA incremental aligner), which does a local alignment of a seed and extends it.

Several identification methods exists:

- Phenotypic analyses

    o   fatty acid analyses

    o   Growth conditions (Agar plate, Biolog multiwell plates)

- Genetic analyses

    o   DNA-DNA hybridization

    o   DNA profiling

    o   Sequence

    o   GC ratios

- Phylogenetic analyses

    o   16S-based phylogeny

    o   phylogeny based on other genes

    o   Multi-gene sequence analysis

    o   Whole-genome sequence based analysis

New Species

The minimal standards for describing a new species depend on which group the species belongs to. *c.f.*

Candidatus

Candidatus is a component of the taxonomic name for a bacterium that cannot be maintained in a Bacteriology Culture Collection. It is an interim taxonomic status for noncultivable organisms. e.g. "Candidatus Pelagibacter ubique"

Species Concept

Bacteria divide asexually and for the most part do not show regionalisms ("Everything is everywhere"), therefore the concept of species, which works best for animals, becomes entirely a matter of judgement.

The number of named species of bacteria and archaea (approximately 13,000) is surprisingly small considering their early evolution, genetic diversity and residence in all ecosystems. The reason for this is the differences in species concepts between the *bacteria* and macro-organisms, the difficulties in growing/characterising in pure culture (a prerequisite to naming new species, *vide supra*) and extensive horizontal gene transfer blurring the distinction of species.

The most commonly accepted definition is the polyphasic species definition, which takes into account both phenotypic and genetic differences. However, a quicker diagnostic *ad hoc* threshold to separate species is less than 70% DNA–DNA hybridisation, which corresponds to less than 97% 16S DNA sequence identity. It has been noted that if this were applied to animal classification, the order primates would be a single species.

Pathology vs. Phylogeny

Ideally, taxonomic classification should reflect the evolutionary history of the taxa, i.e. the phylogeny. Although some exceptions are present when the phenotype differs amongst the group, especially from a medical standpoint. Some examples of problematic classifications follow.

Escherichia Coli: Overly Large and Polyphyletic

In the Enterobacteriaceae family of the class Gammaproteobacteria, the species in the genus *Shigella* (*S. dysenteriae, S. flexneri, S. boydii, S. sonnei*) by an evolutionary point of view are strains of the species *Escherichia coli* (polyphyletic), but due to genetic differences cause different medical conditions in the case of the pathogenic strains., Escherichia coli is a badly classified species as some strains share only 20% of their genome. Being so diverse it should be given a higher taxonomic ranking. However, due to the medical conditions associated with the species, it will not be changed to avoid confusion in medical context.

Bacillus Cereus Group: Close and Polyphyletic

In a similar way, the *Bacillus* species (=phylum *Firmicutes*) belonging to the "*B. cereus* group" (*B. anthracis*, *B. cereus*, *B . thuringiensis*, *B. mycoides*, *B. pseudomycoides*, *B. weihenstephanensis* and *B. medusa*) have 99-100% similar 16S rRNA sequence (97% is a commonly cited adequate species cut-off) and are polyphyletic, but for medical reasons (anthrax *etc.*) remain separate.

Yersinia Pestis: Extremely Recent Species

Yersinia pestis is in effect a strain of *Yersinia pseudotuberculosis*, but with a pathogenicity island that confer a drastically different pathology (Black plague and tuberculosis-like symptoms respectively) which arose 15,000 to 20,000 years ago.

Nested Genera in Pseudomonas

In the gammaproteobacterial order *Pseudomonadales*, the genus *Azotobacter* and the species *Azomonas macrocytogenes* are actually members of the genus *Pseudomonas*, but were misclassified due to nitrogen fixing capabilities and the large size of the genus *Pseudomonas* which renders classification problematic. This will probably rectified in the close future.

Nested Genera in Bacillus

Another example of a large genus with nested genera is the *Bacillus* genus, in which the genera *Paenibacillus* and *Brevibacillus* are nested clades. There is insufficient genomic data at present to fully and effectively correct taxonomic errors in *Bacillus*.

Agrobacterium: Resistance to Name Change

Based on molecular data it was shown that the genus *Agrobacterium* is nested in *Rhizobium* and the *Agrobacterium* species transferred to the *Rhizobium* genus (resulting in the following comp. nov.: *Rhizobium radiobacter* (formerly known as *A. tumefaciens*), *R. rhizogenes*, *R. rubi*, *R. undicola* and *R. vitis*) Given the plant pathogenic nature of *Agrobacterium* species, it was proposed to maintain the genus *Agrobacterium* and the latter was counter-argued

Nomenclature

Taxonomic names are written in italics (or underlined when handwritten) with a majuscule first letter with the exception of epithets for species and subspecies. Despite it being common in zoology, tautonyms (e.g. *Bison bison*) are not acceptable and names of taxa used in zoology, botany or mycology cannot be reused for Bacteria (Botany and Zoology do share names).

Nomenclature is the set of rules and conventions which govern the names of taxa. The difference in nomenclature between the various kingdoms/domains is reviewed in.

For Bacteria, valid names must have a Latin or Neolatin name and can only use basic latin letters, consequent-ly hyphens, accents and other letters are not accepted and should be transliterated cor-rectly (e.g. ß=ss). Ancient Greek being written in the Greek alphabet, needs to be trans-literated into the Latin alphabet.

When compound words are created, a connecting vowel is needed depending on the origin of the preceding word, regardless of the word that follows, unless the latter starts with a vowel in which case no connecting vowel is added. If the first compound is Latin then the connecting vowel is an -i-, whereas if the first compound is Greek, the connect-ing vowel is an -o-.

For etymologies of names consult LPSN.

Rules for Higher Taxa

For the *Prokaryotes* (*Bacteria* and *Archaea*) the rank kingdom is not used (although some authors refer to phyla as kingdoms)

If a new or amended species is placed in new ranks, according to Rule 9 of the Bacteri-ologocal Code the name is formed by the addition of an appropriate suffix to the stem of the name of the type genus. For subclass and class the recommendation from is gener-ally followed, resulting in a neutral plural, however a few names do not follow this and instead keep into account graeco-latin grammar (e.g. the female plurals *Thermotogae*, *Aquificae* and *Chlamydiae*, the male plurals *Chloroflexi*, *Bacilli* and *Deinococci* and the greek plurals *Spirochaetes*, *Gemmatimonadetes* and *Chrysiogenetes*).

Rank	Suffix	Example
Genus		*Elusimicrobium*
Subtribe (disused)	-inae	(*Elusimicrobiinae*)
Tribe (disused)	-inae	(*Elusimicrobiieae*)
Subfamily	-oideae	(*Elusimicrobioideae*)
Family	-aceae	*Elusimicrobiaceae*
Suborder	-ineae	(*Elusimicrobineae*)
Order	-ales	*Elusimicrobiales*
Subclass	-idae	(*Elusimicrobidae*)
Class	-ia	*Elusimicrobia*
Phylum	text	*Elusimicrobia*

Phyla Endings

Phyla are not covered by the Bacteriological code, however, the scientific community generally follows the Ncbi and Lpsn taxonomy, where the name of the phylum is gen-erally the plural of the type genus, with the exception of the *Firmicutes, Cyanobacteria* and *Proteobacteria*, whose names do not stem from a genus name. The higher taxa proposed by Cavalier-Smith are generally disregarded by the molecular phylogeny community (*e.g.*) (*vide supra*).

For the *Archaea* the suffix -archaeota is used. For bacterial phyla it was proposed that the suffix -bacteria be used for phyla.

Consequently for main phyla the name of the phyla is the same as the first described class:

- *Acidobacteria* (from *Acidobacterium*)

- *Actinobacteria* (from *Actinomyces*)

- *Caldisericia* (from *Caldisericum*)

- *Elusimicrobia* (from *Elusimicrobium*)

- *Fusobacteria* (from *Fusobacterium*)

- *Thermodesulfobacteria* (from *Thermodesulfobacterium*)

- *Thermotogae* (from *Thermotoga*)

- *Aquificae* (from *Aquifex*)

- *Chlamydiae* (from *Chlamydia*)

- *Chloroflexi* (from *Chloroflexus*)

- *Chrysiogenetes* (from *Chrysiogenes*)

- *Gemmatimonadetes* (from *Gemmatimonas*)

- *Deferribacteres* (from *Deferribacter*)

Whereas for others where the -ia suffix for class is used regardless of grammar they differ:

- phylum *Bacteroidetes* vs. class *Bacteroidia* from *Bacteroides*

- phylum *Chlorobi* vs. class *Chlorobia* from *Chlorobium*

- phylum *Verrucomicrobia* vs. class *Verrucomicrobiae* from *Verrucomicrobium* (anomalous class name)

- phylum *Dictyoglomi* versus class *Dictyoglomia* from *Dictyoglomus*

- phylum *Fibrobacteres* versus class *Fibrobacteria* from *Fibrobacter* (c.f. the suffix -bacter, note the difference with *Deferribacteres*)

- phylum *Lentisphaerae* versus class *Lentisphaeria* from *Lentisphaera*

- phylum *Nitrospira* or *Nitrospirae* versus class *Nitrospira* from *Nitrospira*

- phylum *Spirochaetes* versus class *Spirochaetae* from *Spirochaeta*

- phylum *Synergistetes* versus class *Synergistetia* from *Synergistes*

- phylum *Planctomycetes* versus *Planctomycea* from *Planctomyces*

An exception is the phylum *Deinococcus-Thermus*, which bears a hyphenated pair of genera —only non accented latin letters are accepted for valid names, but phyla are not officially recognised.

Names After People

Many species are named after people, either the discoverer or a famous person in the field of microbiology, for example *Salmonella* is after D.E. Salmon, who discovered it (albeit as "Bacillus typhi").

For the generic epithet, all names derived from people must be in the female nominative case, either by changing the ending to -a or to the diminutive -ella, depending on the name.

For the specific epithet, the names can be converted into either adjectival form (adding -nus (m.), -na (f.), -num (n.) according to the gender of the genus name) or the genitive of the latinised name.

Names After Places

Many species (the specific epithet) are named after the place they are present or found (e.g. Thiospirillum jenense). Their names are created by forming an adjective by joining the locality's name with the ending -ensis (m. or f.) or ense (n.) in agreement with the gender of the genus name, unless a classical Latin adjective exists for the place. However, names of places should not be used as nouns in the genitive case.

Vernacular Names

Despite the fact that some hetero/homogeneus colonies or biofilms of bacteria have names in English (e.g. dental plaque or Star jelly), no bacterial species has a vernacular/trivial/common name in English.

For names in the singular form, plurals cannot be made (singulare tantum) as would imply multiple groups with the same label and not multiple members of that group (by analogy, in English, chairs and tables are types of furniture, which cannot be used in the plural form "furnitures" to describe both members), conversely names plural form are pluralia tantum. However, a partial exception to this is made by the use of vernacular names. However, to avoid repetition of taxonomic names which break the flow of prose, vernacular names of members of a genus or higher taxa are often used and recommended, these are formed by writing the name of the taxa in sentence case roman ("standard" in MS Office) type, therefore treating the proper noun as an English common noun (e.g. the salmonellas), although there is some debate about the grammar of plurals, which can either be regular plural by adding -(e)s (the salmonellas) or using the ancient Greek or Latin plural form (irregular plurals) of the noun (the salmonellae); the latter is problematic as the plural of - bacter would be -bacteres, while the plural of myces (N.L. masc. n. from Gr. masc. n. mukes) is mycetes.

Customs are present for centain names, such as those ending in -monas are converted into -monad (one pseudomonad, two aeromonads and not -monades).

Bacteria which are the etiological cause for a disease are often referred to by the disease name followed by a describing noun (bacterium, bacillus, coccus, agent or the name of their phylum) e.g. cholera bacterium (*Vibrio cholerae*) or Lyme disease spirochete (*Borrelia burgdorferi*) , note also rickettsialpox (*Rickettsia akari*).

Treponema is converted into treponeme and the plural is treponemes and not treponemata.

Some unusual bacteria have special names such as Quin's oval (*Quinella ovalis*) and Walsby's square (*Haloquadratum walsbyi*).

Before the advent of molecular phylogeny, many higher taxonomic groupings had only trivial names, which are still used today, some of which are polyphyletic, such as Rhizobacteria. Some higher taxonomic trivial names are:

- Blue-green algae are members of the phylum *Cyanobacteria*
- Green non-sulfur bacteria are members of the phylum *Chloroflexi*
- Green sulfur bacteria are members of the *Chlorobi*
- Purple bacteria are some, but not all, members of the phylum Proteobacteria
- Purple sulfur bacteria are members of the order *Chromatiales*
- low G+C Gram-positive bacteria are members of the phylum *Firmicutes*, regardless of GC content
- high G+C Gram-positive bacteria are members of the phylum *Actinobacteria*, regardless of GC content
- Rhizobacteria are members of various genera of proteobacteria
- Rhizobia are members of the order *Rhizobiales*
- Lactic streptococci are members of the genus *Lactococcus*
- Coryneform bacteria are members of the family *Corynebacteriaceae*
- Fruiting gliding bacteria or myxobacteria are members of the order *Myxococcales*
- Enterics are members of the order Enterobacteriales, although the term is

avoided if they do not live in the intestines, such as *Pectobacterium*

- Acetic acid bacteria are members of the family *Acetobacteraceae*

Terminology

- The abbreviation for species is sp. (plural spp.) and is used after a generic epithet to indicate a species of that genus. Often used to denote a strain of a genus for which the species is not known either because has the organism has not been described yet as a species or insufficient tests were conducted to identify it. For example *Halomonas* sp. GFAJ-1

- If a bacterium is known and well-studied but not culturable, it is given the term Candidatus in its name

- A basonym is original name of a new combination, namely the first name given to a taxa before it was reclassified

- A synonym is an alternative name for a taxa, i.e. a taxa was erroneusly described twice

- When a taxon is transferred it becomes a new combination (comb. nov.) or nomina nova (nom. nov.)

- paraphyly, monophyly and polyphyly

Monera

Monera was a kingdom that contained unicellular organisms with a prokaryotic cell organization (having no nuclear membrane), such as bacteria. The taxon Monera was first proposed as a phylum by Ernst Haeckel in 1866. Subsequently, the phylum was elevated to the rank of kingdom in 1925 by Édouard Chatton. The last commonly accepted megaclassification with the taxon Monera was the five-kingdom classification system established by Robert Whittaker in 1969.

Under the three-domain system of taxonomy, introduced by Carl Woese in 1977, which reflects the evolutionary history of life, the organisms found in kingdom Monera have been divided into two domains, Archaea and Bacteria (with Eukarya as the third domain). Furthermore, the taxon Monera is paraphyletic (does not include all descendants of their most recent common ancestor), as Archaea and Eukarya are currently believed to be more closely related than either is to Bacteria. The term "moneran" is the informal name of members of this group and is still sometimes used (as is the term "prokaryote") to denote a member of either domain.

Most bacteria were classified under Monera; however, Cyanobacteria (often called the blue-green algae) were initially classified under Plantae due to their ability to photosynthesize.

History

Haeckel's Classification

Traditionally the natural world was classified as animal, vegetable, or mineral as in Systema Naturae. After the development of the microscope, attempts were made to fit microscopic organisms into either the plant or animal kingdoms. In 1676, Antonie van Leeuwenhoek discovered bacteria and called them "animalcules," assigning them to the class Vermes of the Animalia. Due to the limited tools — the sole references for this group were shape, behaviour, and habitat — the description of genera and their classification was extremely limited, which was accentuated by the perceived lack of importance of the group.

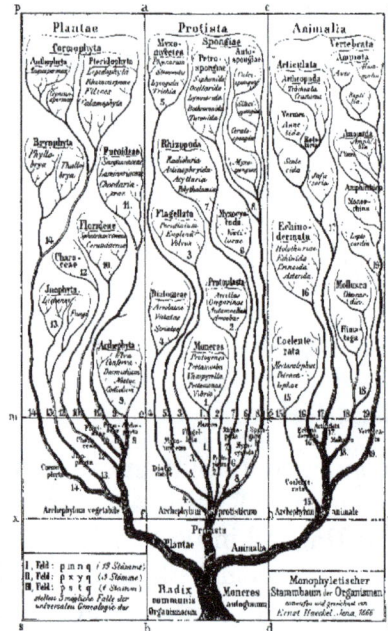

Tree of Life in Generelle Morphologie der Organismen (1866)

Ten years after *The Origin of Species* by Charles Darwin, in 1866 Ernst Haeckel, a supporter of evolution, proposed a three-kingdom system which added the Protista as a new kingdom that contained most microscopic organisms. One of his eight major divisions of Protista was composed of the monerans (called Moneres by Haeckel), which he defined as completely structureless and homogeneous organisms, consisting only of a piece of plasma. Haeckel's Monera included not only bacterial groups of early discovery but also several small eukaryotic organisms; in fact the genus Vibrio is the only bacterial genus explicitly assigned to the phylum, while others are mentioned indirectly, which led Copeland to speculate that Haeckel considered all bacteria to belong to the genus Vibrio, ignoring other bacterial genera. One notable exception were the members of the modern phylum Cyanobacteria, such as *Nostoc*, which were placed in the phylum Archephyta of Algae (vide infra: Blue-green algae).

The Neolatin noun Monera and the German noun Moneren/Moneres are derived from the ancient Greek noun *moneres* () which Haeckel states to mean "simple", however it actually means "single, solitary". Haeckel also describes the protist genus Monas in the two pages about Monera in his 1866 book. The informal name of a member of the Monera was initially moneron, but later moneran was used.

Due to its lack of features, the phylum was not fully subdivided, but the genera therein were divided into two groups:

- die Gymnomoneren (no envelope [sic.]): Gymnomonera

  o *Protogenes* — such as *Protogenes primordialis*, an unidentified amoeba (eukaryote) and not a bacterium

  o *Protamaeba* — an incorrectly described/fabricated species

  o *Vibrio* — a genus of comma-shaped bacteria first described in 1854

  o *Bacterium* — a genus of rod-shaped bacteria first described in 1828. Haeckel does not explicitly assign this genus to the Monera.

  o *Bacillus* — a genus of spore-forming rod-shaped bacteria first described in 1835 Haeckel does not explicitly assign this genus to the Monera kingdom.

  o *Spirochaeta* — thin spiral-shaped bacteria first described in 1835 Haeckel does not explicitly assign this genus to the Monera.

  o *Spirillum* — spiral-shaped bacteria first described in 1832 Haeckel does not explicitly assign this genus to the Monera.

  o *etc.*: Haeckel does provide a comprehensive list.

- die Lepomoneren (with envelope): Lepomonera

  o *Protomonas* — identified to a synonym of *Monas*, a flagellated protozoan, and not a bacterium. The name was reused in 1984 for an unrelated genus of bacteria.

  o *Vampyrella* — now classed as a eukaryote and not a bacterium.

Subsequent Classifications

Like Protista, the Monera classification was not fully followed at first and several different ranks were used and located with animals, plants, protists or fungi.

Furthermore, Häkel's classification lacked specificity and was not exhaustive — it in fact covers only a few pages—, consequently a lot of confusion arose even to the point that the Monera did not contain bacterial genera and others according to Huxley. They were first recognized as a kingdom by Enderlein in 1925 (Bakterien-Cyclogenie. de Gruyter, Berlin).

The most popular scheme was created in 1859 by C. Von Nägeli who classified non-phototrophic Bacteria as the class Schizomycetes.

The class Schizomycetes was then emended by Walter Migula (along with the coinage of the genus *Pseudomonas* in 1894) and others. This term was in dominant use even in 1916 as reported by Robert Earle Buchanan, as it had priority over other terms such as Monera. However, starting with Ferdinand Cohn in 1872 the term *bacteria* (or in German *Bacterien*) became prominently used to informally describe this group of species without a nucleus: Bacterium was in fact a genus created in 1828 by Christian Gottfried Ehrenberg Additionally, Cohn divided the bacteria according to shape namely:

- Spherobacteria for the cocci

- Microbacteria for the short, non-filamentous rods

- Desmobacteria for the longer, filamentous rods and Spirobacteria for the spiral forms.

Successively, Cohn created the Schizophyta of Plants which contained the non-photrophic bacteria in the family Schizomycetes and the phototrophic bacteria (blue green algae/Cyanobacteria) in the Schizophyceae This union of blue green algae and Bacteria was much later followed by Haeckel, who classified the two families in a revised phylum Monera in the Protista.

Stanier and van Neil (1941, The main outlines of bacterial classification. J Bacteriol 42: 437- 466) recognized the Kingdom Monera with 2 phyla, Myxophyta and Schizomycetae, the latter comprising classes Eubacteriae (3 orders), Myxobacteriae (1 order), and Spirochetae (1 order); Bisset (1962, Bacteria, 2nd ed., Livingston, London) distinguished 1 class and 4 orders: Eubacteriales, Actinomycetales, Streptomycetales, and Flexibacteriales; Orla-Jensen (1909, Die Hauptlinien des naturalischen Bakteriensystems nebst einer Ubersicht der Garungsphenomene. Zentr. Bakt. Parasitenk., II, 22: 305-346) and Bergey et al (1925, Bergey's Manual of Determinative Bacteriology, Baltimore : Williams & Wilkins Co.) with many subsequent editions) also presented classifications.

Rise to Prominence

The term Monera became well established in the 20s and 30s when to rightfully increase the importance of the difference between species with a nucleus and without, In

1925 Édouard Chatton divided all living organisms into two empires Prokaryotes and Eukaryotes: the Kingdom Monera being the sole member of the Prokaryotes empire.

The anthropic importance of the crown group of animals, plants and fungi was hard to depose; consequently, several other megaclassification schemes ignored on the empire rank but maintained the kingdom Monera consisting of bacteria, such Copeland in 1938 and Whittaker in 1969. The latter classification system was widely followed, in which Robert Whittaker proposed a five kingdom system for classification of living organisms. Whittaker's system placed most single celled organisms into either the prokaryotic Monera or the eukaryotic Protista. The other three kingdoms in his system were the eukaryotic Fungi, Animalia, and Plantae. Whittaker, however, did not believe that all his kingdoms were monophyletic. Whittaker subdiveded the kingdom into two branches containing several phyla:

- Myxomonera branch

    o Cyanophyta, now called Cyanobacteria

    o Myxobacteria

- Mastigomonera branch

    o Eubacteriae

    o Actinomycota

    o Spirochaetae

Alternative commonly followed subdivision systems were based on Gram stains. This culminated in the Gibbons and Murray classification of 1978:

- Gracilicutes (gram negative)

    o Photobacteria (photosynthetic): class Oxyphotobacteriae (water as electron acceptor, includes the order Cyanobacteriales = blue green algae, now phylum Cyanobacteria) and class Anoxyphotobacteriae (anaerobic phototrophs, orders: Rhodospirillales and Chlorobiales

    o Scotobacteria (non-photosynthetic, now the Proteobacteria and other gram negative nonphotosynthetic phyla)

- Firmacutes [sic] (gram positive, subsequently corrected to Firmicutes)

    o several orders such as Bacillales and Actinomycetales (now in the phylum Actinobacteria)

- Mollicutes (gram variable, e.g. Mycoplasma)

- Mendocutes (uneven gram stain, "methanogenic bacteria" now known as the Archaea)

Three-domain System

In 1977, a PNAS paper by Carl Woese and George Fox demonstrated that the archaea (initially called archaebacteria) are not significantly closer in relationship to the bacteria than they are to eukaryotes. The paper received front-page coverage in *The New York Times*, and great controversy initially. The conclusions have since become accepted, leading to replacement of the kingdom Monera with the two kingdoms Bacteria and Archaea. A minority of scientists, including Thomas Cavalier-Smith, continue to reject the widely accepted division between these two groups. Cavalier-Smith has published classifications in which the archaebacteria are part of a subkingdom of the Kingdom Bacteria.

Blue-green Algae

Although it was generally accepted that one could distinguish prokaryotes from eukaryotes on the basis of the presence of a nucleus, mitosis versus binary fission as a way of reproducing, size, and other traits, the monophyly of the kingdom Monera (or for that matter, whether classification should be according to phylogeny) was controversial for many decades. Although distinguishing between prokaryotes from eukaryotes as a fundamental distinction is often credited to a 1937 paper by Édouard Chatton (little noted until 1962), he did not emphasize this distinction more than other biologists of his era. Roger Stanier and C. B. van Niel believed that the bacteria (a term which at the time did not include blue-green algae) and the blue-green algae had a single origin, a conviction which culminated in Stanier writing in a letter in 1970, "I think it is now quite evident that the blue-green algae are not distinguishable from bacteria by any fundamental feature of their cellular organization". Other researchers, such as E. G. Pringsheim writing in 1949, suspected separate origins for bacteria and blue-green algae. In 1974, the influential *Bergey's Manual* published a new edition coining the term cyanobacteria to refer to what had been called blue-green algae, marking the acceptance of this group within the Monera.

Summary

Linnaeus 1735	Haeckel 1866	Chatton 1925	Copeland 1938	Whittaker 1969	Woese et al. 1990	Cavalier-Smith 1998
2 kingdoms	3 kingdoms	2 empires	4 kingdoms	5 kingdoms	3 domains	6 kingdoms
(not treated)	Protista	Prokaryota	Monera	Monera	Bacteria	Bacteria
					Archaea	Bacteria
		Eukaryota	Protoctista	Protista	Eucarya	Protozoa
						Chromista
Vegetabilia	Plantae		Plantae	Plantae		Plantae
				Fungi		Fungi
Animalia	Animalia		Animalia	Animalia		Animalia

Bacterial Phyla

The bacterial phyla are the major lineages, known as phyla or divisions, of the domain *Bacteria*.

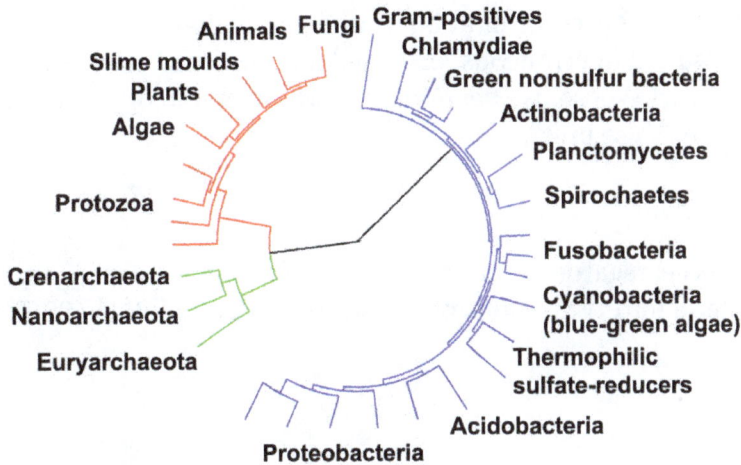

Phylogenetic tree showing the diversity of bacteria, compared to other organisms.
Eukaryotes are colored red, archaea green and bacteria blue.

In the scientific classification established by Carl von Linné, each bacterial strain has to be assigned to a species (binary nomenclature), which is a lower level of a hierarchy of ranks. Currently, the most accepted mega-classification system is under the three-domain system, which is based on molecular phylogeny. In that system, bacteria are members of the domain *Bacteria* and "phylum" is the rank below domain, since the rank "kingdom" is disused at present in bacterial taxonomy. When bacterial nomenclature was controlled under the Botanical Code, the term *division* was used, but now that bacterial nomenclature (with the exception of cyanobacteria) is controlled under the Bacteriological Code, the term *phylum* is preferred.

In this classification scheme, *Bacteria* is (unofficially) subdivided into 30 phyla with representatives cultured in a lab. Many major clades of bacteria that cannot currently be cultured are known solely and somewhat indirectly through metagenomics, the analysis of bulk samples from the environment. If these possible clades, candidate phyla, are included, the number of phyla is 52 or higher. Therefore, the number of major phyla has increased from 12 identifiable lineages in 1987, to 30 in 2014, or over 50 including candidate phyla. The total number has been estimated to exceed 1,000 bacterial phyla.

At the base of the clade Bacteria, close to the last universal common ancestor of all living things, some scientists believe there may be a definite branching order, whereas other scientists, such as Norman Pace, believe there was a large hard polytomy, a simultaneous multiple speciation event.

Molecular Phylogenetics

Traditionally, phylogeny was inferred and taxonomy established based on studies of morphology. Recently molecular phylogenetics has been used to allow better elucidation of the evolutionary relationship of species by analysing their DNA and protein sequences, for example their ribosomal DNA. The lack of easily accessible morphological features, such as those present in animals and plants, hampered early efforts of classification and resulted in erroneous, distorted and confused classification, an example of which, noted Carl Woese, is *Pseudomonas* whose etymology ironically matched its taxonomy, namely "false unit".

Initial Sub-division

In 1987, Carl Woese, regarded as the forerunner of the molecular phylogeny revolution, divided Eubacteria into 11 divisions based on 16S ribosomal RNA (SSU) sequences:

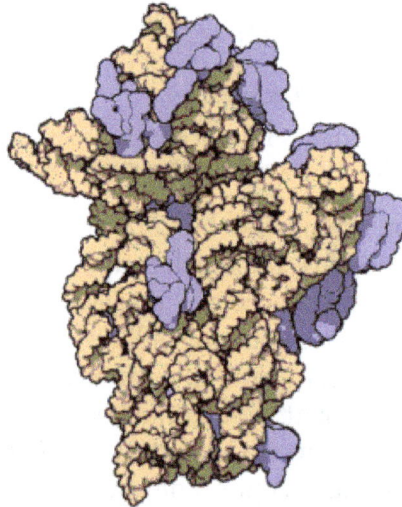

Atomic structure of the 30S ribosomal Subunit from Thermus thermophilus of which 16S makes up a part. Proteins are shown in blue and the single RNA strand in orangeish.

- Purple Bacteria and their relatives (later renamed Proteobacteria)

    o   alpha subdivision (purple non-sulfur bacteria, rhizobacteria, *Agrobacterium, Rickettsiae, Nitrobacter*)

    o   beta subdivision (*Rhodocyclus*, (some) *Thiobacillus, Alcaligenes, Spirillum, Nitrosovibrio*)

    o   gamma subdivision (enterics, fluorescent pseudomonads, purple sulfur bacteria, *Legionella*, (some) *Beggiatoa*)

    o   delta subdivision (Sulfur and sulfate reducers (*Desulfovibrio*), Myxobacteria, *Bdellovibrio*)

- Gram-positive Eubacteria

  o High-G+C species (later renamed Actinobacteria) (*Actinomyces, Strep-tomyces, Arthrobacter, Micrococcus, Bifidobacterium*)

  o Low-G+C species (later renamed Firmicutes) (*Clostridium, Peptococ-cus, Bacillus, Mycoplasma*)

  o Photosynthetic species (*Heliobacterium*)

  o Species with Gram-negative walls (*Megasphaera, Sporomusa*)

- Cyanobacteria and chloroplasts (*Aphanocapsa, Oscillatoria, Nostoc, Synechoc-occus, Gloeobacter, Prochloron*)

- Spirochetes and relatives

  o Spirochetes (*Spirochaeta, Treponema, Borrelia*)

  o Leptospiras (*Leptospira, Leptonema*)

- Green sulfur bacteria (*Chlorobium, Chloroherpeton*)

- Bacteroides, Flavobacteria and relatives (later renamed Bacteroidetes

  o Bacteroides (*Bacteroides, Fusobacterium*)

  o Flavobacterium group (*Flavobacterium, Cytophaga, Saprospira, Flex-ibacter*)

- Planctomyces and relatives (later renamed Planctomycetes)

  o Planctomyces group (*Planctomyces, Pasteuria* [sic])

  o Thermophiles (*Isocystis pallida*)

- Chlamydiae (*Chlamydia psittaci, Chlamydia trachomatis*)

- Radioresistant micrococci and relatives (now commonly referred to as Deino-coccus–Thermus or Thermi)

  o Deinococcus group (*Deinococcus radiodurans*)

  o Thermophiles (*Thermus aquaticus*)

- Green non-sulfur bacteria and relatives (later renamed Chloroflexi)

  o Chloroflexus group (*Chloroflexus, Herpetosiphon*)

  o Thermomicrobium group (*Thermomicrobium roseum*)

- Thermotogae (*Thermotoga maritima*)

New Cultured Phyla

New species have been cultured since 1987, when Woese's review paper was published, that are sufficiently different that they warrant a new phylum. Most of these are thermophiles and often also chemolithoautotrophs, such as Aquificae, which oxidises hydrogen gas. Other non-thermophiles, such as Acidobacteria, a ubiquitous phylum with divergent physiologies, have been found, some of which are chemolithotrophs, such as *Nitrospira* (nitrile-oxidising) or *Leptospirillum* (Fe-oxidising)., some proposed phyla however do not appear in LPSN as they were insufficiently described or are awaiting approval or it is debated if they may belong to a pre-existing phylum. An example of this is the genus *Caldithrix*, consisting of *C. palaeochoryensis* and *C. abyssi*, which is considered Deferribacteres, however, it shares only 81% similarity with the other Deferribacteres (*Deferribacter* species and relatives) and is considered a separate phylum by Rappé and Giovannoni. Additionally the placement of the genus *Geovibrio* in the phylum Deferribacteres is debated.

Uncultivated Phyla and Metagenomics

With the advent of methods to analyse environmental DNA (metagenomics), the 16S rRNA of an extremely large number of undiscovered species have been found, showing that there are several whole phyla which have no known cultivable representative and that some phyla lack in culture major subdivisions as is the case for Verrucomicrobia and Chloroflexi. The term Candidatus is used for proposed species for which the lack of information prevents it to be validated, such as where the only evidence is DNA sequence data, even if the whole genome has been sequenced. When the species are members of a whole phylum it is called a candidate division (or candidate phylum) and in 2003 there were 26 candidate phyla out of 52. A candidate phylum was defined by Hugenholtz and Pace in 1998, as a set of 16S ribosomal RNA sequences with less than 85% similarity to already-described phyla. More recently an even lower threshold of 75% was proposed. Three candidate phyla were known before 1998, prior to the 85% threshold definition by Hugenholtz and Pace:

- OS-K group (from Octopus spring, Yellowstone National Park)

- Marine Group A (from Pacific ocean)

- Termite Group 1 (from a *Reticulitermes speratus* termite gut, now Elusimicrobia)

Since then several other candidate phyla have been identified:)

- OP1, OP3, OP5 (now Caldiserica), OP8, OP9, OP10 (now Armatimonadetes), OP11 (obsidian pool, Yellowstone National Park)

- WS2, WS3, WS5, WS6 (Wurtsmith contaminated aquifer)

- SC3 and SC4 (from arid soil)

- vadinBE97 (now Lentisphaerae)

- NC10 (from flooded caves)

- BRC1 (from bulk soil and rice roots)

- ABY1 (from sediment)

- Guyamas1 (from hydrothermal vents)

- GN01, GN02, GN04 (from a Guerrero Negro hypersaline microbial mat)

- NKB19 (from activated sludge)

- SBR1093 (from activated sludge)

- TM6 and TM7 (Torf, Mittlere Schicht ["peat, middle layer"])

Since then a candidate phylum called Poribacteria was discovered, living in symbiosis with sponges and extensively studied. (Note: the divergence of the major bacterial lineages predates sponges) Another candidate phylum, called Tectomicrobia, was also found living in symbiosis with sponges. And *Nitrospina gracilis*, which had long eluded phylogenetic placement, was proposed to belong to a new phylum, Nitrospinae.

Other candidate phyla that have been the center of some studies are TM7, the genomes of organisms of which have even been sequenced (draft), WS6 and Marine Group A.

Two species of the candidate phylum OP10, which is now called Armatimonadetes, where recently cultured: *Armatimonas rosea* isolated from the rhizoplane of a reed in a lake in Japan and *Chthonomonas calidirosea* from an isolate from geothermally heated soil at Hell's Gate, Tikitere, New Zealand.

One species, *Caldisericum exile*, of the candidate phylum OP5 was cultured, leading to it being named Caldiserica.

The candidate phylum VadinBE97 is now known as Lentisphaerae after *Lentisphaera araneosa* and *Victivallis vadensis* were cultured.

More recently several candidate phyla have been given provisional names despite the fact that they have no cultured representatives:

- Candidate phylum ACD58 was renamed Berkelbacteria

- Candidate phylum CD12 (also known as candidate phylum BHI80-139) was renamed Aerophobetes

- Candidate phylum EM19 was renamed Calescamantes

- Candidate phylum GN02 (also known as candidate phylum BD1-5) was re-named Gracilibacteria

- Candidate phylum KSB3 was renamed Modulibacteria

- Candidate phylum NKB19 was renamed Hydrogenedentes

- Candidate phylum OctSpa1-106 was renamed Fervidibacteria

- Candidate phylum OD1 was renamed Parcubacteria

- Candidate phylum OP1 was renamed Acetothermia

- Candidate phylum OP3 was renamed Omnitrophica

- Candidate phylum OP8 was renamed Aminicenantes

- Candidate phylum OP9 (also known as candidate phylum JS1) was renamed Atribacteria

- Candidate phylum OP11 was renamed Microgenomates

- Candidate phylum PER was renamed Perigrinibacteria

- Candidate phylum SAR406 (also known as candidate phylum Marine Group A) was renamed Marinimicrobia

- Candidate phylum SR1 was renamed Absconditabacteria

- Candidate phylum TM6 was renamed Dependentiae

- Candidate phylum TM7 was renamed Saccharibacteria

- Candidate phylum WS3 was renamed Latescibacteria

- Candidate phylum WWE1 was renamed Cloacimonetes

- Candidate phylum WWE3 was renamed Katanobacteria

- Candidate phylum ZB1 was renamed Ignavibacteriae

Despite these lineages not being officially recognised (due to the ever-increasing number of sequences belonging to undescribed phyla) the ARB-Silva database lists 67 phyla, including 37 candidate phyla (Acetothermia, Aerophobetes, Aminicenantes, aquifer1, aquifer2, Atribacteria, Calescamantes, CKC4, Cloacimonetes, GAL08, GOUTA4, Gracilibacteria, Fermentibacteria (Hyd24-12), Hydrogenedentes, JL-ETNP-Z39, Kazan-3B-09, Latescibacteria, LCP-89, LD1-PA38, Marinimicrobia, Microgenomates, OC31, Omnitrophica, Parcubacteria, PAUC34f, RsaHF231, S2R-29, Saccharibacteria, SBYG-2791, SHA-109, SM2F11, SR1, TA06, TM6, WCHB1-60, WD272, and WS6), the Ribosomal Database Project 10, lists 49 phyla, including 20 candidate phyla (Acetothermia, Aminicenantes, Atribacteria, BRC1, Calescamantes, Cloacimonetes, Hydrogenedentes, Ignavibacteriae, Latescibacteria, Marinimicrobia, Microgenomates, Nitrospinae, Omnitrophica, Parcubacteria,

Poribacteria, SR1, Saccharibacteria, WPS-1, WPS-2, and ZB3), and NCBI lists 120 phyla, including 90 candidate phyla (AC1, Acetothermia, Aerophobetes, Aminicenantes, Atribacteria, Berkelbacteria, BRC1, CAB-I, Calescamantes, CPR1, CPR2, CPR3, EM 3, Fervidibacteria, GAL15, GN01, GN03, GN04, GN05, GN06, GN07, GN08, GN09, GN10, GN11, GN12, GN13, GN14, GN15, Gracilibacteria, Fermentibacteria (Hyd24-12), Hydrogenedentes, JL-ETNP-Z39, KD3-62, kpj58rc, KSA1, KSA2, KSB1, KSB2, KSB3, KSB4, Latescibacteria, marine group A, Marinimicrobia, Microgenomates, MSBL2, MSBL3, MSBL4, MSBL5, MSBL6, NC10, Nitrospinae, NPL-UPA2, NT-B4, Omnitrophica, OP2, OP4, OP6, OP7, OS-K, Parcubacteria, Peregrinibacteria, Poribacteria, RF3, Saccharibacteria, SAM, SBR1093, Sediment-1, Sediment-2, Sediment-3, Sediment-4, SPAM, SR1, TA06, TG2, TM6, VC2, WOR-1, WOR-3, WPS-1, WPS-2, WS1, WS2, WS4, WS5, WS6, WWE3, WYO, ZB3, and Zixibacteria).

Superphyla

Despite the unclear branching order for most bacterial phyla, several groups of phyla have clear clustering and are referred to as superphyla.

The FCB Group

The FCB group includes Bacteroidetes, the unplaced genus *Caldithrix*, Chlorobi, candidate phylum Cloacimonetes, Fibrobacteres, Gemmatimonadates, candidate phylum Ignavibacteriae, candidate phylum Latescibacteria, candidate phylum Marinimicrobia, and candidate phylum Zixibacteria.

The PVC Group

The PVC group includes Chlamydiae, Lentisphaerae, candidate phylum Omnitrophica, Planctomycetes, candidate phylum Poribacteria, and Verrucomicrobia.

Patescibacteria

The proposed superphylum, Patescibacteria, includes candidate phyla Gracilibacteria, Microgenomates, Parcubacteria, and Saccharibacteria and possibly Dependentiae. These same candidate phyla, along with candidate phyla Berkelbacteria, CPR2, CPR3, Kazan, Perigrinibacteria, SM2F11, WS6, and WWE3 were more recently proposed to belong to the larger CPR Group. To complicate matters, it has been suggested that sev-eral of these phyla are themselves actually superphyla.

Terrabacteria

The proposed superphylum, Terrabacteria, includes Actinobacteria, Cyanobacteria, Deinococcus–Thermus, Chloroflexi, Firmicutes, and candidate phylum OP10.

Proteobacteria as Superphylum

It has been proposed that several of the classes of the phylum Proteobacteria are phyla in their own right, which would make Proteobacteria a superphylum.

Cryptic Superphyla

Several candidate phyla (Microgenomates, Omnitrophica, Parcubacteria, and Sacchari-bacteria) and several accepted phyla (Elusimicrobia, Caldiserica, and Armatimonadetes) have been suggested to actually be superphyla that were incorrectly described as phyla because rules for defining a bacterial phylum are lacking. For example, it is suggested that candidate phylum Microgenomates is actually a superphylum that encompasses 28 subordinate phyla and that phylum Elusimocrobia is actually a superphylum that encompasses 7 subordinate phyla.

Overview of Phyla

As of January 2016, there are 30 phyla in the domain "Bacteria" accepted by LPSN. There are no fixed rules to the nomenclature of bacterial phyla. It was proposed that the suffix "-bacteria" be used for phyla, but generally the name of the phylum is generally the plural of the type genus, with the exception of the *Firmicutes*, *Cyanobacteria*, and *Proteobacteria*, whose names do not stem from a genus name (*Actinobacteria* instead is from *Actinomyces*).

Acidobacteria

The *Acidobacteria* (diderm Gram negative) is most abundant bacterial phylum in many soils, but its members are mostly uncultured. Additionally, they phenotypically diverse and include not only acidophiles, but also many non-acidophiles. Generally its members divide slowly, exhibit slow metabolic rates under low-nutrient conditions and can tolerate well fluctuations in soil hydration.

Actinobacteria

The *Actinobacteria* is a phylum of monoderm Gram positive bacteria, many of which notable secondary metabolite producers. There are only two phyla of monoderm Gram positive bacteria, the other being the *Firmicutes*; the actinobacteria generally have higher GC content so are sometimes called "high-CG Gram positive bacteria". Notable genera/species include *Streptomyces* (antibiotic production), *Propioni-bacterium acnes* (odorous skin commensal) and *Propionibacterium freudenreichii* (holes in Emmental)

Aquificae

The *Aquificae* (diderm Gram negative) contains only 14 genera (including Aquifex and Hydrogenobacter). The species are hyperthermophiles and chemolithotrophs (sulphur). According to some studies may be one of the most deep branching phyla.

Armatimonadetes

Bacteroidetes

The *Bacteroidetes* (diderm Gram negative) is a member of the FBC superphylum. Some species are opportunistic pathogens, while other are the most common human gut commensal bacteria. Gained notoriety in the non-scientific community with the urban myth as a bacterial weight loss powder.

Caldiserica

This phylum was formerly known as candidate phylum OP5, *Caldisericum exile* is the sole representative.

Chlamydiae

The *Chlamydiae* (diderms, weakly Gram negative) is a phylum of the PVC superphylum. It is composed of only 6 genera of obligate intracellular pathogens with a complex life cycle. Species include *Chlamydia trachomatis* (chlamydia infection).

Chlorobi

Chlorobi is a member of the FBC superphylum. It contains only 7 genera of obligately anaerobic photoautotrophic bacteria, known colloquially as Green sulfur bacteria. The reaction centre for photosynthesis in Chlorobi and Chloroflexi (another photosynthetic group) is formed by a structures called the chlorosome as opposed to phycobilisomes of cyanobacteria (another photosynthetic group).

Chloroflexi

Chloroflexi,diverse phylum including thermophiles and halorespirers, known colloquially as Green non-sulfur bacteria.

Chrysiogenetes

Chrysiogenetes, only 3 genera (*Chrysiogenes arsenatis, Desulfurispira natronophila, Desulfurispirillum alkaliphilum*)

Cyanobacteria

Cyanobacteria, major photosynthetic clade believed to have caused Earth's oxygen atmosphere, also known as the blue-green algae

Deferribacteres

Deinococcus–Thermus

Deinococcus–Thermus, *Deinococcus radiodurans* and *Thermus aquaticus* are "commonly known" species of this phylum.

Dictyoglomi

Elusimicrobia

Elusimicrobia, formerly candidate phylum Termite Group 1

Fibrobacteres

Fibrobacteres, member of the FBC superphylum.

Firmicutes

Firmicutes, Low-G+C Gram positive species most often spore-forming, in two/three classes: the class Bacilli such as the *Bacillus* spp. (e.g. *B. anthracis*, a pathogen, and *B. subtilis*, biotechnologically useful), lactic acid bacteria (e.g. *Lactobacillus casei* in yoghurt, *Oenococcus oeni* in malolactic fermentation, *Streptococcus pyogenes*, pathogen), the class Clostridia of mostly anaerobic sulphite-reducing saprophytic species, includes the genus *Clostridium* (e.g. the pathogens *C. dificile, C. tetani, C. botulinum* and the biotech *C. acetobutylicum*)

Fusobacteria

Gemmatimonadetes

Gemmatimonadetes, member of the FBC superphylum.

Lentisphaerae

Lentisphaerae, formerly clade VadinBE97, member of the PVC superphylum.

Nitrospirae

Planctomycetes

Planctomycetes, member of the PVC superphylum.

Proteobacteria

Proteobacteria, contains most of the "commonly known" species, such as *Escherichia coli* and *Pseudomonas aeruginosa*.

Spirochaetes

Spirochaetes, notable for compartmentalisation and species include *Borrelia burgdorferi*, which causes Lyme disease.

Synergistetes

The *Synergistetes* is a phylum whose members are diderm Gram negative, rod-shaped obligate anaerobes, some of which are human commensals.

Tenericutes

The *Tenericutes* includes the class *Mollicutes*, formerly/debatedly of the phylum *Firmicutes* (sister clades). Despite their monoderm Gram-positive relatives, they lack peptidoglycan. Notable genus: *Mycoplasma*.

Thermodesulfobacteria

The *Thermodesulfobacteria* is a phylum composed of only three genera in the same family (*Thermodesulfobacteriaceae*: *Caldimicrobium*, *Thermodesulfatator* and *Thermodesulfobacterium*). The members of the phylum are thermophilic sulphate-reducers.

Thermomicrobia

Thermotogae

The *Thermotogae* is a phylum of whose members possess an unusual outer envelope called the toga and are mostly hyperthermophilic obligate anaerobic fermenters.

Verrucomicrobia

Verrucomicrobia is a phylum of the PVC superphylum. Like the *Planctomycetes* species, its members possess a compartmentalised cell plan with a condensed nucleoid and the ribosomes pirellulosome (enclosed by the intracytoplasmic membrane) and paryphoplasm compartment between the intracytoplasmic membrane and cytoplasmic membrane.

Branching Order

The branching order of the phyla of bacteria is unclear. Different studies arrive at different results due to different datasets and methods. For example, in studies using 16S and few other sequences *Thermotogae* and *Aquificae* appear as the most basal phyla, whereas in several phylogenomic studies, *Firmicutes* are the most basal.

- Branching order of bacterial phyla (Woese, 1987)

- Branching order of bacterial phyla (Rappe and Giovanoni, 2004)

- Branching order of bacterial phyla after ARB Silva Living Tree

- Branching order of bacterial phyla (Ciccarelli *et al.*, 2006)

- Branching order of bacterial phyla (Battistuzzi *et al.*, 2004)

- Branching order of bacterial phyla (Gupta, 2001)

- Branching order of bacterial phyla (Cavalier-Smith, 2002)

References

- Madigan, Michael (2009). Brock Biology of Microorganisms. San Francisco: Pearson/Benjamin Cummings. ISBN 0-13-232460-1.

- Ernst Heinrich Philipp August Haeckel (1867). Generelle Morphologie der Organismen. Reimer, Berlin. ISBN 1-144-00186-2.

- Don J. Brenner; Noel R. Krieg; James T. Staley (July 26, 2005) [1984(Williams & Wilkins)]. George M. Garrity, ed. Introductory Essays. Bergey's Manual of Systematic Bacteriology. 2A (2nd ed.). New York: Springer. p. 304. ISBN 978-0-387-24143-2.

- Lapage SP; Sneath PHA; Lessel EF; Skerman VBD; Seeliger HPR; Clark WA, eds. (1992). International Code of Nomenclature of Bacteria, 1990 Revision. Washington (DC): American Society for Microbiology. ISBN 1-55581-039-X.

- Madigan M (2009). Brock Biology of Microorganisms. San Francisco: Pearson/Benjamin Cummings. ISBN 0-13-232460-1.

- Boone DR; Castenholz RW (May 18, 2001) [1984 (Williams & Wilkins)]. Garrity GM, ed. The Archaea and the Deeply Branching and Phototrophic Bacteria. Bergey's Manual of Systematic Bacteriology. 1 (2nd ed.). New York: Springer. p. 721. ISBN 978-0-387-98771-2.

- Euzéby JP. "List of Prokaryotic names with Standing in Nomenclature: Caldithrix". Archived from the original on 22 January 2011. Retrieved 30 December 2010.

Types of Bacteria

The types of bacteria that have been discussed in the chapter are epiphytic bacteria, gram-negative bacteria, gram-positive bacteria, indicator bacteria and cyanobacteria. Epiphytic bacteria live on the surface of a plant; different bacterias prefer different plants to live on. This section explains to the reader the types of bacteria that exist.

Epiphytic Bacteria

Epiphytic bacteria are bacteria which live non-parasitically on the surface of a plant on various organs such as the leaves, roots, flowers, buds, seeds and fruit. In current studies it has been determined that epiphytic bacteria generally don't harm the plant, but promote the formation of ice crystals. Some produce an auxin hormone which promotes plant growth and plays a role in the life cycle of the bacteria.

Different bacteria prefer different plants and different plant organs depending on the organ's nutritional content, and depending on the bacteria's colonization system which is controlled by the host plant. Bacteria which live on leaves are referred to as phyllobacteria, and bacteria which live on the root system are referred to as rhizabacteria. They adhere to the plant surface forms as 1-cluster 2- individual bacterial cell 3- biofilm . The age of the organ also affects the epiphytic bacteria population and characteristics and has a role in the inhibition of phytopathogen on plant. Epiphytic bacteria found in the marine environment have a role in the nitrogen cycle.

Species

There are diverse species of epiphytic bacteria, for example:

Citrobacter youngae

Bacillus thuringiensis

Enterobacter soli

Bacillus tequilensis

Bacillus aryabhattai

Pantoea eucalypti

Pseudomonas palleroniana

Serratia nematodiphila

Stenotrophomonas maltophilia

Pseudomonas mosselii

Pseudomonas putida

Lysinibacillus xylanilyticus

Enterobacter asburiae

Acinetobacter johnsonii

Pseudomonas macerans

Classification

Many epiphytic bacteria are rod-shaped, and classified as either gram negative or gram positive, pigmented or non-pigmented, fermentative or non-fermentative .

Non-pigmented epiphytic bacteria have high a GC content in their genome, a characteristic which protects the bacteria from the ultraviolet rays of the sun. Because of this, these bacteria have special nutritional requirements. Current studies on epiphytic bacteria are underway for biotechnological applications areas such as the promotion of plant growth. Epiphytic bacteria are removed from the plant surface through ultraviolet radiation, chemical surface disinfection, and washing .

Gram-negative Bacteria

Gram-negative bacteria are a group of bacteria that do not retain the crystal violet stain used in the Gram staining method of bacterial differentiation. They are characterized by their cell envelopes, which are composed of a thin peptidoglycan cell wall sandwiched between an inner cytoplasmic cell membrane and a bacterial outer membrane.

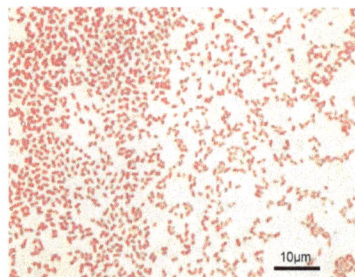

Microscopic image of gram-negative *Pseudomonas aeruginosa* bacteria (pink-red rods)

Gram-negative bacteria are spread worldwide, in virtually all environments that support life. The gram-negative bacteria include the model organism *Escherichia coli*, as well as many pathogenic bacteria, such as *Pseudomonas aeruginosa*, *Neisseria gonorrhoeae*, Chlamydia trachomatis, and *Yersinia pestis*. Several classes of antibiotics target gram-negative bacteria specifically, including aminoglycosides and carbapenems.

Characteristics

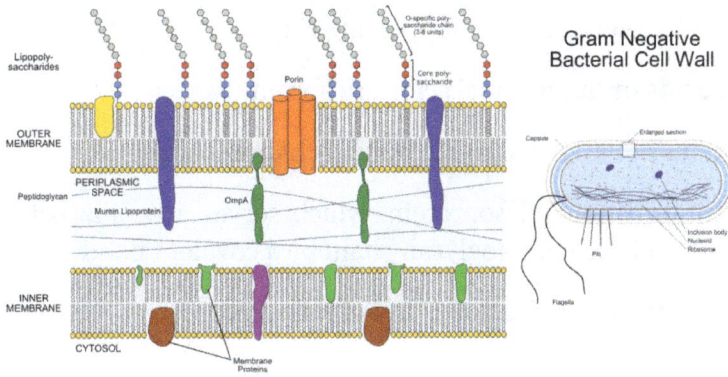

Gram-negative cell wall structure

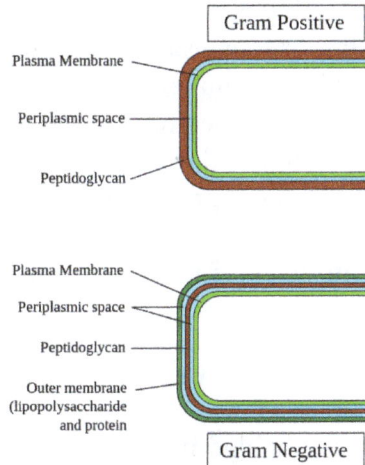

Gram-positive and -negative bacteria are differentiated chiefly by their cell wall structure

Gram-negative bacteria display these characteristics:

1. An inner cell membrane is present (cytoplasmic)

2. A thin peptidoglycan layer is present (This is much thicker in gram-positive bacteria)

3. Has outer membrane containing lipopolysaccharides (LPS, which consists of lipid A, core polysaccharide, and O antigen) in its outer leaflet and phospholipids in the inner leaflet

4. Porins exist in the outer membrane, which act like pores for particular molecules

5. Between the outer membrane and the cytoplasmic membrane there is a space filled with a concentrated gel-like substance called periplasm

6. The S-layer is directly attached to the outer membrane rather than to the peptidoglycan

7. If present, flagella have four supporting rings instead of two

8. Teichoic acids or lipoteichoic acids are absent

9. Lipoproteins are attached to the polysaccharide backbone

10. Some contain Braun's lipoprotein, which serves as a link between the outer membrane and the peptidoglycan chain by a covalent bond

11. Most, with very few exceptions, do not form spores

Classification

Along with cell shape, Gram staining is a rapid diagnostic tool and once was used to group species at the subdivision of Bacteria. Historically, the kingdom Monera was divided into four divisions based on Gram staining: Firmacutes (+), Gracillicutes (−), Mollicutes (0) and Mendocutes (var.). Since 1987, the monophyly of the gram-negative bacteria has been disproven with molecular studies. However some authors, such as Cavalier-Smith still treat them as a monophyletic taxon (though not a clade; his definition of monophyly requires a single common ancestor but does not require holophyly, the property that all descendents be encompassed by the taxon) and refer to the group as subkingdom "Negibacteria".

Outer Cell Membrane Bacterial Classification

Though bacteria are traditionally divided into the two main groups of gram-positive and gram-negative, based on their Gram stain retention, this classification system can be ambiguous; the results of Gram-staining, the cell-envelope organization of the bacteria, and the taxonomic group that the bacteria belongs to, do not necessarily match up for some bacterial species. A positive or negative Gram stain is also not a reliable characteristic for phylogenetic sorting, as these gram-positive and gram-negative bacteria do not form phylogenetically coherent groups. However, Gram staining response of bacteria is an empirical criterion; it is based on the marked differences in the ultrastructure and chemical composition of the two main kinds of prokaryotic cells that are found in nature. These two kinds of cells are distinguished from each other based upon the presence or absence of an outer lipid membrane, which is a reliable and fundamental characteristic of bacterial cells.

The term Monoderm bacteria or Monoderm prokaryotes has been proposed for the bacterial (prokaryotic) cells that are bounded by a single cell membrane. All gram-positive bacteria are bounded by only a single unit lipid membrane and generally contain a thick layer (20–80 nm) of peptidoglycan, which is responsible for retaining the Gram stain. A number of other bacteria that are bounded by a single membrane but stain gram-negative either because they lack the peptidoglycan layer (viz., mycoplasmas) or are unable to retain the Gram stain because of their cell wall composition, also show a close evolutionary relationship to the gram-positive bacteria.

The term Diderm bacteria has been proposed for bacteria/prokaryotes that are bounded by a cytoplasmic membrane and an outer cell membrane. In contrast to gram-positive bacteria, all archetypical gram-negative bacteria have a cytoplasmic membrane as well as an outer cell membrane, and contain only a thin layer of peptidoglycan (2–3 nm) between the membranes. The inner and outer cell membranes form a new compartment in these cells, the periplasmic space or the periplasmic compartment. The distinction between the monoderm and diderm prokaryotes is also supported by conserved signature indels in a number of important proteins (viz. DnaK, GroEL).

Of these two structurally distinct groups of prokaryotic organisms, monoderm prokaryotes are indicated to be ancestral. Based upon a number of different observations including that the gram-positive bacteria are the major reactors to antibiotics and that gram-negative bacteria are, in general, resistant to them, it has been proposed that the outer cell membrane in gram-negative bacteria (diderms) evolved as a protective mechanism against antibiotic selection pressure. Some bacteria such as Deinococcus, which stain gram-positive due to the presence of a thick peptidoglycan layer, but also possess an outer cell membrane are suggested as intermediates in the transition between monoderm (gram-positive) and diderm (gram-negative) bacteria. The diderm bacteria can also be further differentiated between simple diderms lacking lipopolysaccharide; the archetypical diderm bacteria, in which the outer cell membrane contains lipopolysaccharide; and the diderm bacteria, in which the outer cell membrane is made up of mycolic acid (e. g. Mycobacterium).

In addition, a number of bacterial taxa (viz. Negativicutes, Fusobacteria, Synergistetes, and Elusimicrobia) that are either part of the phylum Firmicutes or branches in its proximity are also found to possess a diderm cell structure. However, a conserved signature indel (CSI) in the HSP60 (GroEL) protein distinguishes all traditional phyla of gram-negative bacteria (e.g., Proteobacteria, Aquificae, Chlamydiae, Bacteroidetes, Chlorobi, Cyanobacteria, Fibrobacteres, Verrucomicrobia, Planctomycetes, Spirochetes, Acidobacteria) from these other atypical diderm bacteria as well as other phyla of monoderm bacteria (e.g., Actinobacteria, Firmicutes, Thermotogae, Chloroflexi). The presence of this CSI in all sequenced species of conventional lipopolysaccharide-containing gram-negative bacterial phyla provides evidence that these phyla of bacteria form a monophyletic clade and that no loss of the outer membrane from any species from this group has occurred.

Example Species

The proteobacteria are a major group of gram-negative bacteria, including *Escherichia coli* (*E. coli*), *Salmonella*, *Shigella*, and other Enterobacteriaceae, *Pseudomonas*, *Moraxella*, *Helicobacter*, *Stenotrophomonas*, *Bdellovibrio*, acetic acid bacteria, *Legionella* etc. Other notable groups of gram-negative bacteria include the cyanobacteria, spirochaetes, green sulfur, and green non-sulfur bacteria.

Medically relevant gram-negative cocci include the four types that cause a sexually transmitted disease (*Neisseria gonorrhoeae*), a meningitis (*Neisseria meningitidis*), and respiratory symptoms (*Moraxella catarrhalis*, *Haemophilus influenzae*).

Medically relevant gram-negative bacilli include a multitude of species. Some of them cause primarily respiratory problems (*Klebsiella pneumoniae*, *Legionella pneumophila*, *Pseudomonas aeruginosa*), primarily urinary problems (*Escherichia coli*, *Proteus mirabilis*, *Enterobacter cloacae*, *Serratia marcescens*), and primarily gastrointestinal problems (*Helicobacter pylori*, *Salmonella enteritidis*, *Salmonella typhi*).

Gram-negative bacteria associated with hospital-acquired infections include *Acinetobacter baumannii*, which cause bacteremia, secondary meningitis, and ventilator-associated pneumonia in hospital intensive-care units.

Bacterial Transformation

Transformation is one of three processes for horizontal gene transfer, in which exogenous genetic material passes from bacterium to another, the other two being conjugation (transfer of genetic material between two bacterial cells in direct contact) and transduction (injection of foreign DNA by a bacteriophage virus into the host bacterium). In transformation, the genetic material passes through the intervening medium, and uptake is completely dependent on the recipient bacterium.

As of 2014 about 80 species of bacteria were known to be capable of transformation, about evenly divided between Gram-positive and Gram-negative bacteria; the number might be an overestimate since several of the reports are supported by single papers. Transformation has been studied in medically important Gram-negative bacteria species such as *Helicobacter pylori*, *Legionella pneumophila*, *Neisseria meningitidis*, *Neisseria gonorrhoeae*, *Haemophilus influenzae* and *Vibrio cholerae*. It has also been studied in gram-negative species found in soil such as *Pseudomonas stutzeri*, *Acinetobacter baylyi*, and gram-negative plant pathogens such as *Ralstonia solanacearum* and *Xylella fastidiosa*.

Medical Treatment

One of the several unique characteristics of gram-negative bacteria is the structure of the bacterial outer membrane. The outer leaflet of this membrane comprises a complex lipopolysaccharide (LPS) whose lipid portion acts as an endotoxin. If gram-negative bacteria enter the circulatory system, the liposaccharide can cause a toxic reaction. This results in fever, an increased respiratory rate, and low blood pressure. This may lead to life-threatening condition of endotoxic shock.

The outer membrane protects the bacteria from several antibiotics, dyes, and detergents that would normally damage either the inner membrane or the cell wall (made of peptidoglycan). The outer membrane provides these bacteria with resistance to lysozyme and penicillin. However, alternative medicinal treatments such as lysozyme with EDTA and the antibiotic ampicillin have been developed to combat the protective outer membrane of some pathogenic gram-negative organisms. Other drugs can also be used, significant ones being chloramphenicol, streptomycin, and nalidixic acid. Chloramphenicol is rarely used in the EU due to the association with drug induced pancytopenia.

The pathogenic capability of gram-negative bacteria is often associated with certain components of their membrane, in particular, the LPS. In humans, the presence of LPS triggers an innate immune response, activating the immune system and producing cytokines (hormonal regulators). Inflammation is a common reaction to cytokine production, which can also produce host toxicity. The innate immune response to LPS, however, is not synonymous with pathogenicity, or the ability to cause disease.

Orthographic Note

The adjectives 'gram-positive' and 'gram-negative' derive from the surname of Hans Christian Gram; as eponymous adjectives, their initial letter can be either lowercase 'g' or capital 'G', depending on whose style guide (if any) governs the document being written. This is further explained at *Gram staining § Orthographic note.*

Types of Gram-negative Bacteria

Escherichia Coli

Escherichia coli is a gram-negative, facultatively anaerobic, rod-shaped non-spore-forming rod, ~1-2 μm wide, 3-30 μm long</microbiologybytes.com> bacterium of the genus *Escherichia* that is commonly found in the lower intestine of warm-blooded organisms (endotherms). Most *E. coli* strains are harmless, but some serotypes can cause serious food poisoning in their hosts, and are occasionally responsible for product recalls due to food contamination. The harmless strains are part of the normal flora of the gut, and can benefit their hosts by producing vitamin K_2, and preventing colonization of the intestine with pathogenic bacteria. *E. coli* is expelled into the environment within fecal matter. The bacterium grows massively in fresh fecal matter under aerobic conditions for 3 days, but its numbers decline slowly afterwards.

E. coli and other facultative anaerobes constitute about 0.1% of gut flora, and fecal–oral transmission is the major route through which pathogenic strains of the bacterium cause disease. Cells are able to survive outside the body for a limited amount of time, which makes them potential indicator organisms to test environmental samples for fecal contamination. A growing body of research, though, has examined environmentally persistent *E. coli* which can survive for extended periods outside of a host.

The bacterium can be grown and cultured easily and inexpensively in a laboratory setting, and has been intensively investigated for over 60 years. *E. coli* is a chemoheterotroph whose chemically defined medium must include a source of carbon and energy. *E. coli* is the most widely studied prokaryotic model organism, and an important species in the fields of biotechnology and microbiology, where it has served as the host organism for the majority of work with recombinant DNA. Under favorable conditions, it takes only 20 minutes to reproduce.

Biology and Biochemistry

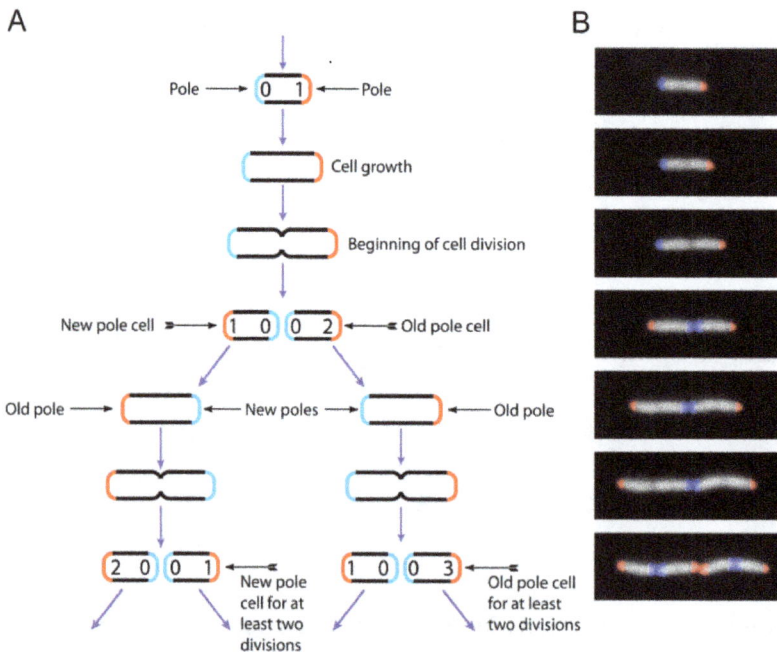

Model of successive binary fission in *E. coli*

A colony of *E. coli* growing

Type and Morphology

E. coli is a gram-negative, facultative anaerobic (that makes ATP by aerobic respiration if oxygen is present, but is capable of switching to fermentation or anaerobic respiration if oxygen is absent) and nonsporulating bacterium. Cells are typically rod-shaped, and are about 2.0 micrometers (μm) long and 0.25–1.0 μm in diameter, with a cell volume of 0.6–0.7 μm^3.

E. coli stains gram-negative because its cell wall is composed of a thin peptidoglycan layer and an outer membrane. During the staining process, *E. coli* picks up the color of the counterstain safranin and stains pink. The outer membrane surrounding the cell wall provides a barrier to certain antibiotics such that *E. coli* is not damaged by penicillin.

Strains that possess flagella are motile. The flagella have a peritrichous arrangement.

Metabolism

E. coli can live on a wide variety of substrates and uses mixed-acid fermentation in anaerobic conditions, producing lactate, succinate, ethanol, acetate, and carbon dioxide. Since many pathways in mixed-acid fermentation produce hydrogen gas, these pathways require the levels of hydrogen to be low, as is the case when *E. coli* lives together with hydrogen-consuming organisms, such as methanogens or sulphate-reducing bacteria.

Culture Growth

Optimum growth of *E. coli* occurs at 37 °C (98.6 °F), but some laboratory strains can multiply at temperatures of up to 49 °C (120.2 °F). *E. coli* grows in a variety of defined laboratory media, such as lysogeny broth, or any medium that contains glucose, ammonium phosphate, monobasic, sodium chloride, magnesium sulfate, potassium phosphate, dibasic, and water. Growth can be driven by aerobic or anaerobic respiration, using a large variety of redox pairs, including the oxidation of pyruvic acid, formic acid, hydrogen, and amino acids, and the reduction of substrates such as oxygen, nitrate, fumarate, dimethyl sulfoxide, and trimethylamine N-oxide. *E. coli* is classified as a facultative anaerobe. It uses oxygen when it is present and available. It can however, continue to grow in the absence of oxygen using fermentation or anaerobic respiration. The ability to be able to continue growing in the absence of oxygen is an advantage to bacteria because their survival is increased in environments where water predominates.

Cell Cycle

The bacterial cell cycle is divided into three stages. The B period occurs between the completion of cell division and the beginning of DNA replication. The C period encompasses the time it takes to replicate the chromosomal DNA. The D period refers to the stage between the conclusion of DNA replication and the end of cell division. The doubling rate of *E. coli* is higher when more nutrients are available. However, the length of the C and D periods do not change, even when the doubling time becomes less than the sum of the C and D periods. At the fastest growth rates, replication begins before the previous round of replication has completed, resulting in multiple replication forks along the DNA and overlapping cell cycles.

Unlike eukaryotes, prokaryotes do not rely upon either changes in gene expression or changes in protein synthesis to control the cell cycle. This probably explains why they do not have similar proteins to those used by eukaryotes to control their cell cycle, such as cdk1. This has led to research on what the control mechanism is in prokaryotes. Recent evidence suggests that it may be membrane- or lipid-based.

Genetic Adaptation

E. coli and related bacteria possess the ability to transfer DNA via bacterial conjugation or transduction, which allows genetic material to spread horizontally through an existing population. The process of transduction, which uses the bacterial virus called a bacteriophage, is where the spread of the gene encoding for the Shiga toxin from the *Shigella* bacteria to *E. coli* helped produce *E. coli* O157:H7, the Shiga toxin producing strain of *E. coli*.

Gene Nomenclature

Genes in *E. coli* are usually named by 4-letter acronyms that derive from their function (when known). For instance, recA is named after its role in homologous recombination plus the letter A. Functionally related genes are named recB, recC, recD etc. The proteins are named by uppercase acronyms, e.g. RecA, RecB, etc. When the genome of *E. coli* was sequenced, all genes were numbered (more or less) in their order on the genome and abbreviated by b numbers, such as b2819 (=recD) etc. The "b" names were created after Fred Blattner who led the genome sequence effort. Another numbering system was introduced with the sequence of another *E. coli* strain, W3110, which was sequenced in Japan and hence uses numbers starting by JW... (Japanese W3110), e.g. JW2787 (= recD). Hence, recD = b2819 = JW2787. Note, however, that most databases have their own numbering system, e.g. the EcoGene database uses EG10826 for recD. Finally, ECK numbers are specifically used for alleles in the MG1655 strain of *E. coli* K-12. Complete lists of genes and their synonyms can be obtained from databases such as EcoGene or Uniprot.

Proteomics

Proteome

Several studies have investigated the proteome of *E. coli*. By 2006, 1,627 (38%) of the 4,237 open reading frames (ORFs) had been identified experimentally.

Interactome

The interactome of *E. coli* has been studied by affinity purification and mass spectrometry (AP/MS) and by analyzing the binary interactions among its proteins.

Protein complexes. A 2006 study purified 4,339 proteins from cultures of strain K-12 and found interacting partners for 2,667 proteins, many of which had unknown functions at the time. A 2009 study found 5,993 interactions between proteins of the same *E. coli* strain, though these data showed little overlap with those of the 2006 publication.

Binary interactions. Rajagopala *et al.* (2014) have carried out systematic yeast two-hybrid screens with most *E. coli* proteins, and found a total of 2,234 protein-protein interactions. This study also integrated genetic interactions and protein structures and mapped 458 interactions within 227 protein complexes.

Normal Microbiota

E. coli belongs to a group of bacteria informally known as coliforms that are found in the gastrointestinal tract of warm-blooded animals. *E. coli* normally colonizes an infant's gastrointestinal tract within 40 hours of birth, arriving with food or water or from the individuals handling the child. In the bowel, *E. coli* adheres to the mucus of the large intestine. It is the primary facultative anaerobe of the human gastrointestinal tract. (Facultative anaerobes are organisms that can grow in either the presence or absence of oxygen.) As long as these bacteria do not acquire genetic elements encoding for virulence factors, they remain benign commensals.

Therapeutic Use

Nonpathogenic *E. coli* strain Nissle 1917, also known as Mutaflor, and *E. coli* O83:K24:H31 (known as Colinfant) are used as probiotic agents in medicine, mainly for the treatment of various gastroenterological diseases, including inflammatory bowel disease.

Role in Disease

Most *E. coli* strains do not cause disease, but virulent strains can cause gastroenteritis, urinary tract infections, and neonatal meningitis. It can also be characterized by severe abdominal cramps, diarrhea that typically turns bloody within 24 hours, and sometimes fever. In rarer cases, virulent strains are also responsible for bowel necrosis (tissue death) and perforation without progressing to hemolytic-uremic syndrome, peritonitis, mastitis, septicemia, and gram-negative pneumonia.

There is one strain, *E.coli* #0157:H7, that produces the Shiga toxin (classified as a bioterrorism agent). This toxin causes premature destruction of the red blood cells, which then clog the body's filtering system, the kidneys, causing hemolytic-uremic syndrome (HUS). This in turn causes strokes due to small clots of blood which lodge in capillaries in the brain. This causes the body parts controlled by this region of the brain not to work properly. In addition, this strain causes the buildup of fluid (since the kidneys do not work), leading to edema around the lungs and legs and arms. This increase in fluid buildup especially around the lungs impedes the functioning of the heart, causing an increase in blood pressure.

Uropathogenic *E. coli* (UPEC) is one of the main causes of urinary tract infections. It is part of the normal flora in the gut and can be introduced in many ways. In particular for females, the direction of wiping after defecation (wiping back to front) can lead to fecal contamination of the urogenital orifices. Anal intercourse can also introduce this bacterium into the male urethra, and in switching from anal to vaginal intercourse, the male can also introduce UPEC to the female urogenital system.

In May 2011, one *E. coli* strain, O104:H4, was the subject of a bacterial outbreak that began in Germany. Certain strains of *E. coli* are a major cause of foodborne illness. The outbreak started when several people in Germany were infected with enterohemorrhagic *E. coli* (EHEC) bacteria, leading to hemolytic-uremic syndrome (HUS), a medical emergency that requires urgent treatment. The outbreak did not only concern Germany, but also 11 other countries, including regions in North America. On 30 June 2011, the German *Bundesinstitut für Risikobewertung (BfR)* (Federal Institute for Risk Assessment, a federal institute within the German Federal Ministry of Food, Agriculture and Consumer Protection) announced that seeds of fenugreek from Egypt were likely the cause of the EHEC outbreak.

Treatment

The mainstay of treatment is the assessment of dehydration and replacement of fluid and electrolytes. Administration of antibiotics has been shown to shorten the course of illness and duration of excretion of enterotoxigenic *E. coli* (ETEC) in adults in endemic areas and in traveller's diarrhoea, though the rate of resistance to commonly used antibiotics is increasing and they are generally not recommended. The antibiotic used depends upon susceptibility patterns in the particular geographical region. Currently, the antibiotics of choice are fluoroquinolones or azithromycin, with an emerging role for rifaximin. Oral rifaximin, a semisynthetic rifamycin derivative, is an effective and well-tolerated antibacterial for the management of adults with non-invasive traveller's diarrhoea. Rifaximin was significantly more effective than placebo and no less effective than ciprofloxacin in reducing the duration of diarrhoea. While rifaximin is effective in patients with *E. coli*-predominant traveller's diarrhoea, it appears ineffective in patients infected with inflammatory or invasive enteropathogens.

Prevention

ETEC is the type of *E. coli* that most vaccine development efforts are focused on. Antibodies against the LT and major CFs of ETEC provide protection against LT-producing ETEC expressing homologous CFs. Oral inactivated vaccines consisting of toxin antigen and whole cells, i.e. the licensed recombinant cholera B subunit (rCTB)-WC cholera vaccine Dukoral have been developed. There are currently no licensed vaccines for ETEC, though several are in various stages of development. In different trials, the rCTB-WC cholera vaccine provided high (85–100%) short-term protection. An oral ETEC vaccine candidate consisting of rCTB and formalin inactivated *E. coli* bacteria expressing major CFs has been shown in clinical trials to be safe, immunogenic, and effective against severe diarrhoea in American travelers but not against ETEC diarrhoea in young children in Egypt. A modified ETEC vaccine consisting of recombinant *E. coli* strains over expressing the major CFs and a more LT-like hybrid toxoid called LCTBA, are undergoing clinical testing.

Other proven prevention methods for *E. coli* transmission include handwashing and improved sanitation and drinking water, as transmission occurs through fecal contamination of food and water supplies.

Causes and Risk Factors

- Working around livestock
- Consuming unpasteurized dairy product
- Eating undercooked meat
- Drinking impure water

Model Organism in Life Science Research

Role in Biotechnology

Because of its long history of laboratory culture and ease of manipulation, *E. coli* plays an important role in modern biological engineering and industrial microbiology. The work of Stanley Norman Cohen and Herbert Boyer in *E. coli*, using plasmids and restriction enzymes to create recombinant DNA, became a foundation of biotechnology.

E. coli is a very versatile host for the production of heterologous proteins, and various protein expression systems have been developed which allow the production of recombinant proteins in *E. coli*. Researchers can introduce genes into the microbes using plasmids which permit high level expression of protein, and such protein may be mass-produced in industrial fermentation processes. One of the first useful applications of recombinant DNA technology was the manipulation of *E. coli* to produce human insulin.

Many proteins previously thought difficult or impossible to be expressed in *E. coli* in folded form have been successfully expressed in *E. coli*. For example, proteins with multiple disulphide bonds may be produced in the periplasmic space or in the cytoplasm of mutants rendered sufficiently oxidizing to allow disulphide-bonds to form, while proteins requiring post-translational modification such as glycosylation for stability or function have been expressed using the N-linked glycosylation system of *Campylobacter jejuni* engineered into *E. coli*.

Modified *E. coli* cells have been used in vaccine development, bioremediation, production of biofuels, lighting, and production of immobilised enzymes.

Model Organism

E. coli is frequently used as a model organism in microbiology studies. Cultivated strains (e.g. *E. coli* K12) are well-adapted to the laboratory environment, and, unlike wild-type strains, have lost their ability to thrive in the intestine. Many laboratory strains lose their ability to form biofilms. These features protect wild-type strains from antibodies and other chemical attacks, but require a large expenditure of energy and material resources.

In 1946, Joshua Lederberg and Edward Tatum first described the phenomenon known as bacterial conjugation using *E. coli* as a model bacterium, and it remains the primary model to study conjugation. *E. coli* was an integral part of the first experiments to understand phage genetics, and early researchers, such as Seymour Benzer, used *E. coli* and phage T4 to understand the topography of gene structure. Prior to Benzer's research, it was not known whether the gene was a linear structure, or if it had a branching pattern.

E. coli was one of the first organisms to have its genome sequenced; the complete genome of *E. coli* K12 was published by *Science* in 1997.

By evaluating the possible combination of nanotechnologies with landscape ecology, complex habitat landscapes can be generated with details at the nanoscale. On such synthetic ecosystems, evolutionary experiments with *E. coli* have been performed to study the spatial biophysics of adaptation in an island biogeography on-chip.

Studies are also being performed attempting to program *E. coli* to solve complicated mathematics problems, such as the Hamiltonian path problem.

History

In 1885, the German-Austrian pediatrician Theodor Escherich discovered this organism in the feces of healthy individuals. He called it *Bacterium coli commune* because it is found in the colon. Early classifications of prokaryotes placed these in a handful of genera based on their shape and motility (at that time Ernst Haeckel's classification of bacteria in the kingdom Monera was in place).

Bacterium coli was the type species of the now invalid genus *Bacterium* when it was revealed that the former type species ("*Bacterium triloculare*") was missing. Following a revision of *Bacterium*, it was reclassified as *Bacillus coli* by Migula in 1895 and later reclassified in the newly created genus *Escherichia*, named after its original discoverer.

Klebsiella

Klebsiella is a genus of nonmotile, Gram-negative, oxidase-negative, rod-shaped bacteria with a prominent polysaccharide-based capsule. It is named after the German microbiologist Edwin Klebs (1834–1913).

Klebsiella species are found everywhere in nature. This is thought to be due to distinct sublineages developing specific niche adaptations, with associated biochemical adaptations which make them better suited to a particular environment. They can be found in water, soil, plants, insects, animals, and humans.

Klebsiella is named after Edwin Klebs, a German microbiologist in the late nineteenth century. Carl Friedlander described *Klebsiella* bacillus which is why it was termed Friedlander bacillus for many years. The members of the genus *Klebsiella* are a part of the human and animal's normal flora in the nose, mouth and intestines. The species of *Klebsiella* are all gram-negative and non-motile. They tend to be shorter and thicker when compared to others in the *Enterobacteriaceae* family. The cells are rods in shape and generally measures 0.3 to 1.5 µm wide by 0.5 to 5.0 µm long. They can be found singly, in pairs, in chains or linked end to end. *Klebsiella* can grow on ordinary lab medium and do not have special growth requirements, like the other members of *Enterobacteriaceae*. The species are aerobic but facultatively anaerobic. Their ideal growth temperature is 35° to 37°, while their ideal pH level is about 7.2.

List of Species of the Genus *Klebsiella*

- *K. granulomatis*
- *K. oxytoca*
- *K. michiganensis*
- *K. pneumoniae* (type-species)
  - *K. p.* subsp. *ozaenae*
  - *K. p.* subsp. *pneumoniae*
  - *K. p.* subsp. *rhinoscleromatis*
- *K. quasipneumoniae*
  - *K. q.* subsp. *quasipneumoniae*

o *K. q.* subsp. *similipneumoniae*

- *K. variicola*

Features

Klebsiella bacteria tend to be rounder and thicker than other members of the Entero-bacteriaceae family. They typically occur as straight rods with rounded or slightly point-ed ends. They can be found singly, in pairs, or in short chains. Diplobacillary forms are commonly found *in vivo*.

They have no specific growth requirements and grow well on standard laboratory me-dia, but grow best between 35 and 37 °C and at pH 7.2. The species are facultative anaerobes, and most strains can survive with citrate and glucose as their sole carbon sources and ammonia as their sole nitrogen source.

Members of the genus produce a prominent capsule, or slime layer, which can be used for serologic identification, but molecular serotyping may replace this method.

Klebsiella in Humans

Klebsiella species are routinely found in the human nose, mouth, and gastrointestinal tract as normal flora; however, they can also behave as opportunistic human patho-gens. *Klebsiella* species are known to also infect a variety of other animals, both as normal flora and opportunistic pathogens.

Klebsiella organisms can lead to a wide range of disease states, notably pneumonia, urinary tract infections, septicemia, meningitis, diarrhea, and soft tissue infections. *Klebsiella* species have also been implicated in the pathogenesis of ankylosing spon-dylitis and other spondyloarthropathies. The majority of human *Klebsiella* infections are caused by *K. pneumoniae*, followed by *K. oxytoca*. Infections are more common in the very young, very old, and those with other underlying diseases, such as cancer, and most infections involve contamination of an invasive medical device.

During the last 40 years, many trials for constructing effective *K. pneumoniae* vaccines have been tried. Currently, no *Klebsiella* vaccine has been licensed for use in the US. *K. pneumoniae* is the most common cause of nosocomial respiratory tract and premature intensive care infections, and the second-most frequent cause of Gram-negative bacte-raemia and urinary tract infections. Drug-resistant isolates remain an important hos-pital-acquired bacterial pathogen, add significantly to hospital stays, and are especially problematic in high-impact medical areas such as intensive care units. This antimicro-bial resistance is thought to be attributable mainly to multidrug efflux pumps. The abil-ity of *K. pneumoniae* to colonize the hospital environment, including carpeting, sinks, flowers, and various surfaces, as well as the skin of patients and hospital staff, has been identified as a major factor in the spread of hospital-acquired infections.

Klebsiella in Animals

In addition to certain *Klebsiella* spp. being discovered as human pathogens, others such as *K. variicola* have been identified as emerging pathogens in humans and animals alike. For instance, *K. variicola* has been identified as one of the causes of bovine mastitis.

Klebsiella in Plants

In plant systems, *Klebsiella* can be found in a variety of plant hosts. *K. pneumoniae* and *K. oxytoca* are able to fix atmospheric nitrogen into a form that can be used by plants, thus are called associative nitrogen fixers or diazotrophs. The bacteria attach strongly to root hairs and less strongly to the surface of the zone of elongation and the root cap mucilage. They are bacteria of interest in an agricultural context, due to their ability to increase crop yields under agricultural conditions. Their high numbers in plants are thought to be at least partly attributable to their lack of a flagellum, as flagella are known to induce plant defenses. Additionally, *K. variicola* is known to associate with a number of different plants including banana trees, sugarcane and has been isolated from the fungal gardens of leaf-cutter ants.

Pseudomonas

Pseudomonas is a genus of Gram-negative, aerobic Gammaproteobacteria, belonging to the family Pseudomonadaceae and containing 191 validly described species. The members of the genus demonstrate a great deal of metabolic diversity and consequently are able to colonize a wide range of niches. Their ease of culture *in vitro* and availability of an increasing number of *Pseudomonas* strain genome sequences has made the genus an excellent focus for scientific research; the best studied species include *P. aeruginosa* in its role as an opportunistic human pathogen, the plant pathogen *P. syringae*, the soil bacterium *P. putida*, and the plant growth-promoting *P. fluorescens*.

Because of their widespread occurrence in water and plant seeds such as dicots, the pseudomonads were observed early in the history of microbiology. The generic name *Pseudomonas* created for these organisms was defined in rather vague terms by Walter Migula in 1894 and 1900 as a genus of Gram-negative, rod-shaped and polar-flagellated bacteria with some sporulating species, the latter statement was later proved incorrect and was due to refractive granules of reserve materials. Despite the vague description, the type species, *Pseudomonas pyocyanea* (basonym of *Pseudomonas aeruginosa*), proved the best descriptor.

Classification History

Like most bacterial genera, the pseudomonad last common ancestor lived hundreds of millions of years ago. They were initially classified at the end of the 19th century when first identified by Walter Migula. The etymology of the name was not specified at the time and first appeared in the seventh edition of *Bergey's Manual of Systematic Bacteriology* (the main authority in bacterial nomenclature) as Greek *pseudes* "false" and *-monas* "a single unit", which can mean false unit; however, Migula possibly intended it as false *Monas*, a nanoflagellated protist (subsequently, the term "monad" was used in the early history of microbiology to denote unicellular organisms). Soon, other species matching Migula's somewhat vague original description were isolated from many natural niches and, at the time, many were assigned to the genus. However, many strains have since been reclassified, based on more recent methodology and use of approaches involving studies of conservative macromolecules.

Recently, 16S rRNA sequence analysis has redefined the taxonomy of many bacterial species. As a result, the genus *Pseudomonas* includes strains formerly classified in the genera *Chryseomonas* and *Flavimonas*. Other strains previously classified in the genus *Pseudomonas* are now classified in the genera *Burkholderia* and *Ralstonia*.

In 2000, the complete genome sequence of a *Pseudomonas* species was determined; more recently, the sequence of other strains has been determined, including *P. aeruginosa* strains PAO1 (2000), *P. putida* KT2440 (2002), *P. protegens* Pf-5 (2005), *P. syringae* pathovar tomato DC3000 (2003), *P. syringae* pathovar syringae B728a (2005), *P. syringae* pathovar phaseolica 1448A (2005), *P. fluorescens* Pf0-1, and *P. entomophila* L48.

Characteristics

Members of the genus display these defining characteristics:

- Rod-shaped

- Gram-negative

- One or more polar flagella, providing motility

- Aerobic

- Non-spore forming

- Catalase-positive

- Oxidase-positive

Other characteristics that tend to be associated with *Pseudomonas* species (with some exceptions) include secretion of pyoverdine, a fluorescent yellow-green siderophore under iron-limiting conditions. Certain *Pseudomonas* species may also produce additional types of siderophore, such as pyocyanin by *Pseudomonas aeruginosa* and thioquinolobactin by *Pseudomonas fluorescens,*. *Pseudomonas* species also typically give a positive result to the oxidase test, the absence of gas formation from glucose, glucose is oxidised in oxidation/fermentation test using Hugh and Leifson O/F test, beta hemolytic (on blood agar), indole negative, methyl red negative, Voges–Proskauer test negative, and citrate positive.

Pseudomonas may be the most common nucleator of ice crystals in clouds, thereby being of utmost importance to the formation of snow and rain around the world.

Biofilm Formation

All species and strains of *Pseudomonas* have historically been classified as strict aerobes. Exceptions to this classification have recently been discovered in *Pseudomonas* biofilms. A significant number of cells can produce exopolysaccharides associated with biofilm formation. Secretion of exopolysaccharides such as alginate makes it difficult for pseudomonads to be phagocytosed by mammalian white blood cells. Exopolysaccharide production also contributes to surface-colonising biofilms that are difficult to remove from food preparation surfaces. Growth of pseudomonads on spoiling foods can generate a "fruity" odor.

Antibiotic Resistance

Being Gram-negative bacteria, most *Pseudomonas* spp. are naturally resistant to penicillin and the majority of related beta-lactam antibiotics, but a number are sensitive to piperacillin, imipenem, ticarcillin, or ciprofloxacin. Aminoglycosides such as tobramycin, gentamicin, and amikacin are other choices for therapy.

This ability to thrive in harsh conditions is a result of their hardy cell walls that contain porins. Their resistance to most antibiotics is attributed to efflux pumps, which pump out some antibiotics before they are able to act.

Pseudomonas aeruginosa is increasingly recognized as an emerging opportunistic pathogen of clinical relevance. One of its most worrying characteristics is its low antibiotic susceptibility. This low susceptibility is attributable to a concerted action of multidrug efflux pumps with chromosomally encoded antibiotic resistance genes (e.g., *mexAB-oprM*, *mexXY*, etc.,) and the low permeability of the bacterial cellular envelopes. Besides intrinsic resistance, *P. aeruginosa* easily develops acquired resistance either by mutation in chromosomally encoded genes or by the horizontal gene transfer of antibiotic resistance determinants. Development of multidrug resistance by *P. aeruginosa* isolates requires several different genetic events that include acquisition of different mutations and/or horizontal transfer of antibiotic resistance genes. Hypermutation favours the selection of mutation-driven antibiotic resistance in *P. aeruginosa* strains producing chronic infections, whereas the clustering of several different antibiotic resistance genes in integrons favours the concerted acquisition of antibiotic resistance determinants. Some recent studies have shown phenotypic resistance associated to biofilm formation or to the emergence of small-colony-variants may be important in the response of *P. aeruginosa* populations to antibiotic treatment.

Sensitivity to Gallium

Although gallium has no natural function in biology, gallium ions interact with cellular processes in a manner similar to iron(III). When gallium ions are mistakenly taken up in place of iron(III) by bacteria such as Pseudomonas, the ions interfere with respiration, and the bacteria die. This happens because iron is redox-active, allowing the transfer of electrons during respiration, while gallium is redox-inactive.

Taxonomy

The studies on the taxonomy of this complicated genus groped their way in the dark while following the classical procedures developed for the description and identification of the organisms involved in sanitary bacteriology during the first decades of the 20th century. This situation sharply changed with the proposal to introduce as the central criterion the similarities in the composition and sequences of macromolecular components of the ribosomal RNA. The new methodology clearly showed the genus *Pseudomonas*, as classically defined, consists of a conglomerate of genera that could clearly be separated into five so-called rRNA homology groups. Moreover, the taxonomic studies suggested an approach that might prove useful in taxonomic studies of all other prokaryotic groups. A few decades after the proposal of the new genus *Pseudomonas* by Migula in 1894, the accumulation of species names assigned to the genus reached alarming proportions. The number of species in the current list has contracted more than 90%. In fact, this approximated reduction may be even more dramatic if one considers the present list contains many new names; i.e., relatively few names of the original list survived in the process. The new methodology and the inclusion of approaches based on the studies of conservative macromolecules other than rRNA components constitutes an effective prescription that helped to reduce *Pseudomonas* nomenclatural hypertrophy to a manageable size.

Pathogenicity

Animal Pathogens

Infectious species include *P. aeruginosa*, *P. oryzihabitans*, and *P. plecoglossicida*. *P. aeruginosa* flourishes in hospital environments, and is a particular problem in this environment, since it is the second-most common infection in hospitalized patients (nosocomial infections). This pathogenesis may in part be due to the proteins secreted by *P. aeruginosa*. The bacterium possesses a wide range of secretion systems, which export numerous proteins relevant to the pathogenesis of clinical strains.

Plant Pathogens

P. syringae is a prolific plant pathogen. It exists as over 50 different pathovars, many of which demonstrate a high degree of host-plant specificity. Numerous other

Pseudomonas species can act as plant pathogens, notably all of the other members of the *P. syringae* subgroup, but *P. syringae* is the most widespread and best-studied.

Although not strictly a plant pathogen, *P. tolaasii* can be a major agricultural problem, as it can cause bacterial blotch of cultivated mushrooms. Similarly, *P. agarici* can cause drippy gill in cultivated mushrooms.

Use as Biocontrol Agents

Since the mid-1980s, certain members of the *Pseudomonas* genus have been applied to cereal seeds or applied directly to soils as a way of preventing the growth or establishment of crop pathogens. This practice is generically referred to as biocontrol. The biocontrol properties of *P. fluorescens* and *P. protegens* strains (CHA0 or Pf-5 for example) are currently best-understood, although it is not clear exactly how the plant growth-promoting properties of *P. fluorescens* are achieved. Theories include: the bacteria might induce systemic resistance in the host plant, so it can better resist attack by a true pathogen; the bacteria might outcompete other (pathogenic) soil microbes, e.g. by siderophores giving a competitive advantage at scavenging for iron; the bacteria might produce compounds antagonistic to other soil microbes, such as phenazine-type antibiotics or hydrogen cyanide. Experimental evidence supports all of these theories.

Other notable *Pseudomonas* species with biocontrol properties include *P. chlororaphis*, which produces a phenazine-type antibiotic active agent against certain fungal plant pathogens, and the closely related species *P. aurantiaca*, which produces di-2,4-diacetylfluoroglucylmethane, a compound antibiotically active against Gram-positive organisms.

Use as Bioremediation Agents

Some members of the genus are able to metabolise chemical pollutants in the environment, and as a result, can be used for bioremediation. Notable species demonstrated as suitable for use as bioremediation agents include:

- *P. alcaligenes*, which can degrade polycyclic aromatic hydrocarbons.

- *P. mendocina*, which is able to degrade toluene.

- *P. pseudoalcaligenes*, which is able to use cyanide as a nitrogen source.

- *P. resinovorans*, which can degrade carbazole.

- *P. veronii*, which has been shown to degrade a variety of simple aromatic organic compounds.

- *P. putida*, which has the ability to degrade organic solvents such as toluene. At least one strain of this bacterium is able to convert morphine in aqueous solution into the stronger and somewhat expensive to manufacture drug hydromorphone (Dilaudid).

- Strain KC of *P. stutzeri*, which is able to degrade carbon tetrachloride.

Food Spoilage Agents

As a result of their metabolic diversity, ability to grow at low temperatures, and ubiquitous nature, many *Pseudomonas* species can cause food spoilage. Notable examples include dairy spoilage by *P. fragi*, mustiness in eggs caused by *P. taetrolens* and *P. mudicolens*, and *P. lundensis*, which causes spoilage of milk, cheese, meat, and fish.

Species previously classified in the genus

Recently, 16S rRNA sequence analysis redefined the taxonomy of many bacterial species previously classified as being in the *Pseudomonas* genus. Species that moved from the *Pseudomonas* genus are listed below; clicking on a species will show its new classification. The term 'pseudomonad' does not apply strictly to just the *Pseudomonas* genus, and can be used to also include previous members such as the genera *Burkholderia* and *Ralstonia*.

α proteobacteria: *P. abikonensis*, *P. aminovorans*, *P. azotocolligans*, *P. carboxydohydrogena*, *P. carboxidovorans*, *P. compransoris*, *P. diminuta*, *P. echinoides*, *P. extorquens*, *P. lindneri*, *P. mesophilica*, *P. paucimobilis*, *P. radiora*, *P. rhodos*, *P. riboflavina*, *P. rosea*, *P. vesicularis*.

β proteobacteria: *P. acidovorans*, *P. alliicola*, *P. antimicrobica*, *P. avenae*, *P. butanovorae*, *P. caryophylli*, *P. cattleyae*, *P. cepacia*, *P. cocovenenans*, *P. delafieldii*, *P. facilis*, *P. flava*, *P. gladioli*, *P. glathei*, *P. glumae*, *P. graminis*, *P. huttiensis*, *P. indigofera*, *P. lanceolata*, *P. lemoignei*, *P. mallei*, *P. mephitica*, *P. mixta*, *P. palleronii*, *P. phenazinium*, *P. pickettii*, *P. plantarii*, *P. pseudoflava*, *P. pseudomallei*, *P. pyrrocinia*, *P. rubrilineans*, *P. rubrisubalbicans*, *P. saccharophila*, *P. solanacearum*, *P. spinosa*, *P. syzygii*, *P. taeniospiralis*, *P. terrigena*, *P. testosteroni*.

γ-β proteobacteria: *P. beteli*, *P. boreopolis*, *P. cissicola*, *P. geniculata*, *P. hibiscicola*, *P. maltophilia*, *P. pictorum*.

γ proteobacteria: *P. beijerinckii*, *P. diminuta*, *P. doudoroffii*, *P. elongata*, *P. flectens*, *P. halodurans*, *P. halophila*, *P. iners*, *P. marina*, *P. nautica*, *P. nigrifaciens*, *P. pavonacea*, *P. piscicida*, *P. stanieri*.

δ proteobacteria: *P. formicans*.

Bacteriophage

There are a number of bacteriophage that infect *Pseudomonas*, e.g.

- Pseudomonas phage Φ6
- Pseudomonas aeruginosa phage EL
- Pseudomonas aeruginosa phage ΦKMV
- Pseudomonas aeruginosa phage LKD16
- Pseudomonas aeruginosa phage LKA1
- Pseudomonas aeruginosa phage LUZ19
- Pseudomonas aeruginosa phage ΦKZ

Enterobacter

Enterobacter is a genus of common Gram-negative, facultatively anaerobic, rod-shaped, non-spore-forming bacteria of the family Enterobacteriaceae. Several strains of these bacteria are pathogenic and cause opportunistic infections in immunocompromised (usually hospitalized) hosts and in those who are on mechanical ventilation. The urinary and respiratory tracts are the most common sites of infection. The genus *Enterobacter* is a member of the coliform group of bacteria. It does not belong to the fecal coliforms (or thermotolerant coliforms) group of bacteria, unlike *Escherichia coli*, because it is incapable of growth at 44.5 °C in the presence of bile salts. Some of them showed quorum sensing properties as reported before

Two clinically important species from this genus are *E. aerogenes* and *E. cloacae*.

Biochemical Characteristics

The genus *Enterobacter* ferments lactose with gas production during a 48-hour incubation at 35-37 °C in the presence of bile salts and detergents. It is oxidase-negative, indole-negative, and urease-variable.

Treatment

- Note: Treatment is dependent on local trends of antibiotic resistance.
1. Cefepime, a fourth-generation cephalosporin from the β-Lactam antibiotic class.
2. Imipenem (carbapenems) is often the antibiotic of choice.
3. Aminoglycosides such as amikacin have been found to be very effective, as well.
4. Quinolones can be an effective alternative.

Linked to Obesity

A recent study has shown that the presence of *Enterobacter cloacae* B29 in the gut of a morbidly obese individual may have contributed to the patient's obesity. Reduction of the bacterial load within the patient's gut, from 35% *E. cloacae* B29 to non-detectable levels, was associated with a parallel reduction in endotoxin load in the patient and a concomitant, significant reduction in weight. Furthermore, the same bacterial strain, isolated from the patient, induced obesity and insulin resistance in germfree C57BL/6J mice that were being fed a high-fat diet. The study concludes that *E. cloacae* B29 may contribute to obesity in its human hosts through an

Helicobacter

Helicobacter is a genus of Gram-negative bacteria possessing a characteristic helical shape. They were initially considered to be members of the *Campylobacter* genus, but in 1989, Goodwin *et al.* published sufficient reasons to justify the new genus name *Helicobacter*. The *Helicobacter* genus contains about 35 species.

Some species have been found living in the lining of the upper gastrointestinal tract, as well as the liver of mammals and some birds. The most widely known species of the genus is *H. pylori*, which infects up to 50% of the human population. It also serves as the type species of the genus. Some strains of this bacterium are pathogenic to humans, as they are strongly associated with peptic ulcers, chronic gastritis, duodenitis, and stomach cancer.

Helicobacter species are able to thrive in the very acidic mammalian stomach by producing large quantities of the enzyme urease, which locally raises the pH from about 2 to a more biocompatible range of 6 to 7. Bacteria belonging to this genus are usually susceptible to antibiotics such as penicillin, are microaerophilic (optimal oxygen concentration between 5 and 14%) capnophiles, and are fast-moving with their flagella.

Molecular Signatures

Comparative genomic analysis has led to the identification of 11 proteins which are uniquely found in the Helicobacteraceae. Of these proteins, seven are found in all species of the family, while the remaining four are not found in any *Helicobacter* strains and are unique to *Wollinella*. Additionally, a rare genetic event has led to the fusion of the *rpoB* and *rpoC* genes in this family, which is characteristic of them.

Non-helicobacter Pylori Species

Recently, new gastric (*H. suis* and *H. baculiformis*) and enterohepatic (*H. equorum*) species have been reported. *H. pylori* is of primary importance for medicine, but non-*H. pylori* species, which naturally inhabit mammals (except humans) and birds, have been detected in human clinical specimens. These encompass two (gastric and enterohepatic) groups, showing different organ specificity. Importantly, some species, such as *H. hepaticus*, *H. mustelae*, and probably *H. bilis*, exhibit carcinogenic potential in animals. They harbour many virulence genes and may cause diseases not only in animals, but also in humans. Gastric species such as *H. suis* (most often), *H. felis, H. bizzozeronii*, and *H. salomonis* have been associated with chronic gastritis and peptic ulcers in humans and, importantly, with higher risk for MALT lymphoma compared to *H. pylori*. Enterohepatic species e.g., *H. hepaticus, H. bilis*, and *H. ganmani*, have been detected by PCR, but still are not isolated from specimens of patients with hepatobiliary diseases. Moreover, they may be associated with Crohn's disease and ulcerative colitis. The significance of avian helicobacters (*H. pullorum, H. anseris*, and *H. brantae)* also has been evaluated extensively. *H. cinaedi* and *H. canis* can cause severe infections, mostly in immunocompromised patients with animal exposure. Briefly, the role of these species in veterinary and human medicine is increasingly recognised. Several other topics such as isolation of still uncultured species, antibiotic resistance and treatment regimens for infections and pathogenesis and possible carcinogenesis in humans should be evaluated.

Campylobacter

Campylobacter (meaning "curved bacteria") is a genus of Gram-negative bacteria. *Campylobacter* typically appear comma or s-shaped and motile.

Most *Campylobacter* species can cause disease and can infect humans and other animals. The bacterium's main reservoir is poultry; humans can contract the disease from eating food contaminated with *Campylobacter* species. Another source of infection is contact with infected animals, which often carry *Campylobacter* asymptomatically. At least a dozen species of *Campylobacter* have been implicated in human disease, with *C. jejuni* and *C. coli* being the most common. *C. jejuni* is now recognized as one of the main causes of bacterial foodborne disease in many developed countries. *C. jejuni* infection can also spread to the blood in individuals with AIDS, while *C. lari* is a known cause of recurrent diarrhea in children. *C. fetus* is a cause of spontaneous abortions in cattle and sheep, as well as an opportunistic pathogen in humans.

Description

Campylobacter generally appear curved or comma-shaped, and are able to move via unipolar or bipolar flagella. They generally survive in environments with low oxygen. They are positive by the oxidase test and catalase test. *Campylobacter* are nonfermentative.

History

The symptoms of *Campylobacter* infections were described in 1886 in infants by Theodor Escherich. These infections were named cholera infantum, or summer complaint. The genus was first described in 1963; however, the organism was not isolated until 1972.

Genetics

The genomes of several *Campylobacter* species have been sequenced, beginning with *C. jejuni* in 2000. These genome studies have identified molecular markers specific to members of *Campylobacter*. Additionally, several markers were found in all *Campylobacter* species except for *C. fetus*, the most distantly-related species. Many markers were also found which were conserved only between *C. jejuni* and *C. coli*, indicating a close relationship between these two species.

Similar studies have investigated the genes responsible for motility in *Campylobacter* species. All *Campylobacter* species contain two flagellin genes in tandem for motility, *flaA* and *flaB*. These genes undergo intergenic recombination, further contributing to their virulence.

Bacteriophage

The confusing taxonomy of *Campylobacter* over the past decades make it difficult to identify the earliest reports of *Campylobacter* bacteriophages. Bacteriophages specific to the species now known as *C. coli* and *C. fetus* (previously *Vibrio coli* and *V. fetus*), were isolated from cattle and pigs during the 1960s.

Pathogenesis

Campylobacter can cause a gastrointestinal infection called campylobacteriosis. This is characterized by inflammatory, sometimes bloody diarrhea or dysentery syndrome, mostly including cramps, fever, and pain. The most common routes of transmission are fecal-oral, ingestion of contaminated food or water, and the eating of raw meat. Foods implicated in campylobacteriosis include raw or under-cooked poultry, raw dairy products, and contaminated produce. *Campylobacter* is sensitive to the stomach's normal production of hydrochloric acid: as a result, the infectious dose is relatively high, and the bacteria rarely cause illness when a person is exposed to less 10,000 organisms. Nevertheless, people taking antacid medication (e. g. people with gastritis or stomach ulcers) are at higher risk of contracting disease from a smaller amount of organisms, since this type of medication inhibits normal gastric acid. The infection is usually self-limiting and, in most cases, symptomatic treatment by liquid and electrolyte replacement is enough in human infections. The use of antibiotics, though, is controversial. Symptoms typically last five to seven days.

The sites of tissue injury include the jejunum, the ileum, and the colon. Most strains of *C jejuni* produce a toxin (cytolethal distending toxin) that hinders the cells from dividing and activating the immune system. This helps the bacteria to evade the immune system and survive for a limited time in the cells. A cholera-like enterotoxin was once thought to be also made, but this appears not to be the case. The organism produces diffuse, bloody, edematous, and exudative enteritis. Although rarely has the infection been considered a cause of hemolytic uremic syndrome and thrombotic thrombocy-topenic purpura, no unequivocal case reports exist. In some cases, a *Campylobacter* infection can be the underlying cause of Guillain–Barré syndrome. Gastrointestinal perforation is a rare complication of ileal infection.

Campylobacter has also been associated with periodontitis.

Treatment

Diagnosis of campylobacteriosis is made by testing a specimen of feces. Standard treat-ment is azithromycin, a macrolide antibiotic, especially for *Campylobacter* infections in children, although other antibiotics, such as macrolides, quinolones, and tetracy-cline are sometimes used to treat gastrointestinal *Campylobacter* infections in adults. In case of systemic infections, other bactericidal antibiotics are used, such as ampicil-lin, amoxicillin/clavulanic acid, or aminoglycosides. Fluoroquinolone antibiotics, such as ciprofloxacin or levofloxacin, may no longer be effective in some cases due to resis-tance. In addition to antibiotics, dehydrated children may require intravenous fluid treatment in a hospital.

Epidemiology

UK

In January 2013, the UK's Food Standards Agency warned that two-thirds of all raw chicken bought from UK shops was contaminated with *Campylobacter*, affecting an estimated half a million people annually and killing about 100. In June 2014, the Food Standards Agency started a campaign against washing raw chicken, as washing can spread germs by splashing. In May 2015, cumulative results for sam-ples taken from fresh chickens between February 2014 and February 2015 were published by the FSA and showed 73% of chickens tested positive for the presence of *Campylobacter*.

USA

Campylobacter infections increased 14% in the United States in 2012 compared to the period from 2006 to 2008. This represents the highest reported number of infections since 2000.

High prevalence of *Campylobacter* (40% or more) has been reported in raw chick-en meat in retail stores in the USA. The reported prevalence in retail chicken meat is higher than the reported prevalence by the microbiology performance standard testing collected by the U. S. Department of Agriculture, and the last quarterly progress report on *Salmonella* and *Campylobacter* testing of meat and poultry for July–September 2014, published by the Food Safety and Inspection Service of the U. S. Department of Agriculture, shows a low prevalence of *Campylobacter* spp. in ground chicken meat, but a larger prevalence (20%) in mechanically separated chicken meat (which is sold only for further processing).

Canada

FoodNet Canada has reported that *Campylobacter* was the most common pathogen found on packaged chicken breast, with nearly half of all samples testing positive. Additionally, *Campylobacter* and *Salmonella* were the most common causes of gastrointestinal illness in Canada

New Zealand

In August 2016, an estimated 4,000+ residents of Havelock North, a town with 13,000 or so residents, had gastric illness after the water supply was thought to be contaminated by *Campylobacter*.

Gram-positive Bacteria

Gram-positive bacteria are bacteria that give a positive result in the Gram stain test. Gram-positive bacteria take up the crystal violet stain used in the test, and then appear to be purple-coloured when seen through a microscope. This is because the thick peptidoglycan layer in the bacterial cell wall retains the stain after it is washed away from the rest of the sample, in the decolorization stage of the test.

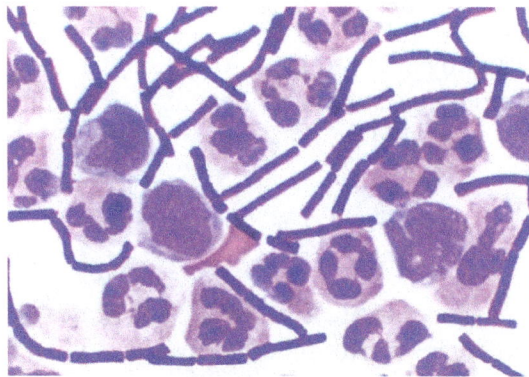

Rod-shaped gram-positive *Bacillus anthracis* bacteria in a cerebrospinal fluid sample stand out from round white blood cells.

Gram-negative bacteria cannot retain the violet stain after the decolorization step; alcohol used in this stage degrades the outer membrane of gram-negative cells making the cell wall more porous and incapable of retaining the crystal violet stain. Their peptidoglycan layer is much thinner and sandwiched between an inner cell membrane and a bacterial outer membrane, causing them to take up the counterstain (safranin or fuchsine) and appear red or pink.

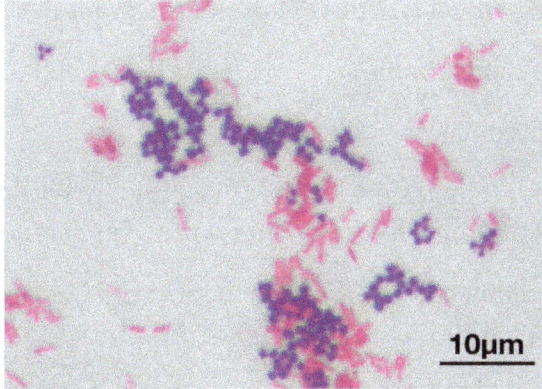

Violet stained gram-positive cocci and pink stained gram-negative rod-shaped bacteria

Despite their thicker peptidoglycan layer, gram-positive bacteria are more receptive to antibiotics than gram-negative, due to the absence of the outer membrane.

Characteristics

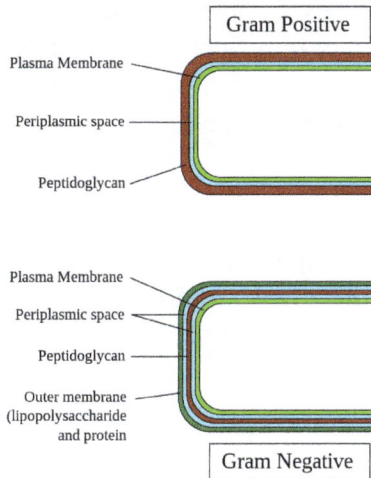

Gram-positive and -negative cell wall structure

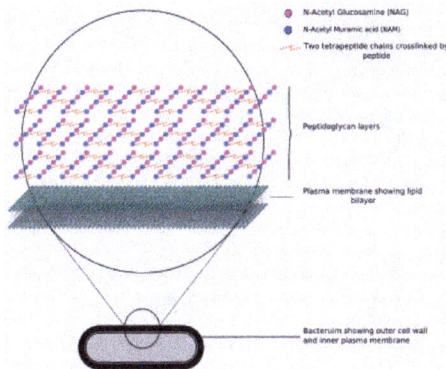

Structure of gram-positive cell wall

In general, the following characteristics are present in gram-positive bacteria:

1. Cytoplasmic lipid membrane

2. Thick peptidoglycan layer

3. Teichoic acids and lipoids are present, forming lipoteichoic acids, which serve as chelating agents, and also for certain types of adherence.

4. Peptidoglycan chains are cross-linked to form rigid cell walls by a bacterial enzyme DD-transpeptidase.

5. A much smaller volume of periplasm than that in gram-negative bacteria.

Only some species have a capsule usually consisting of polysaccharides. Also only some species are flagellates, and when they do have flagella they only have two basal body rings to support them (gram-negative have four). Both gram-positive and gram-negative bacteria commonly have a surface layer called an S-layer. In gram-positive bacteria, the S-layer is attached to the peptidoglycan layer (in gram-negative bacteria, the S-layer is attached directly to the outer membrane). Specific to gram-positive bacteria is the presence of teichoic acids in the cell wall. Some of these are lipoteichoic acids, which have a lipid component in the cell membrane that can assist in anchoring the peptidoglycan.

Classification

Along with cell shape, Gram staining is a rapid method used to differentiate bacterial species. Such staining, together with growth requirement and antibiotic susceptibility testing, and other macroscopic and physiologic tests, forms the full basis for classification and sub-division of the bacteria.

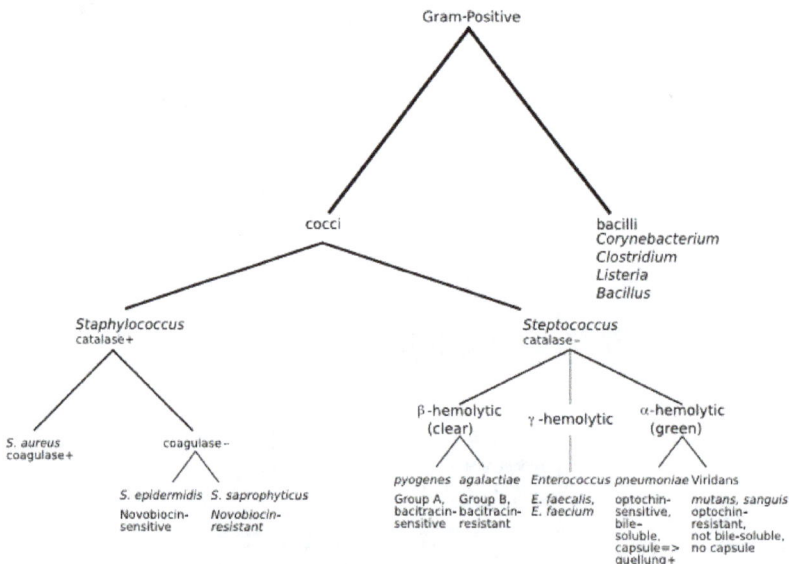

Species identification hierarchy in clinical settings

Historically, the kingdom Monera was divided into four divisions based primarily on Gram staining: Firmicutes (positive in staining), Gracilicutes (negative in staining), Mollicutes (neutral in staining) and Mendocutes (variable in staining). Based on 16S ribosomal RNA phylogenetic studies of the late microbiologist Carl Woese and collaborators and colleagues at the University of Illinois, the monophyly of the gram-positive bacteria was challenged, with major implications for the therapeutic and general study of these organisms. Based on molecular studies of the 16S sequences, Woese recognised twelve bacterial phyla. Two of these were both gram-positive and were divided on the proportion of the guanine and cytosine content in their DNA. The high G + C phylum was made up of the Actinobacteria and the low G + C phylum contained the Firmicutes. The Actinobacteria include the *Corynebacterium*, *Mycobacterium*, *Nocardia* and *Streptomyces* genera. The (low G + C) Firmicutes, have a 45–60% GC content, but this is lower than that of the Actinobacteria.

Importance of the Outer Cell Membrane in Bacterial Classification

Although bacteria are traditionally divided into two main groups, gram-positive and gram-negative, based on their Gram stain retention property, this classification system is ambiguous as it refers to three distinct aspects (staining result, envelope organization, taxonomic group), which do not necessarily coalesce for some bacterial species. The gram-positive and gram-negative staining response is also not a reliable characteristic as these two kinds of bacteria do not form phylogenetic coherent groups. However, although Gram staining response is an empirical criterion, its basis lies in the marked differences in the ultrastructure and chemical composition of the bacterial cell wall, marked by the absence or presence of an outer lipid membrane.

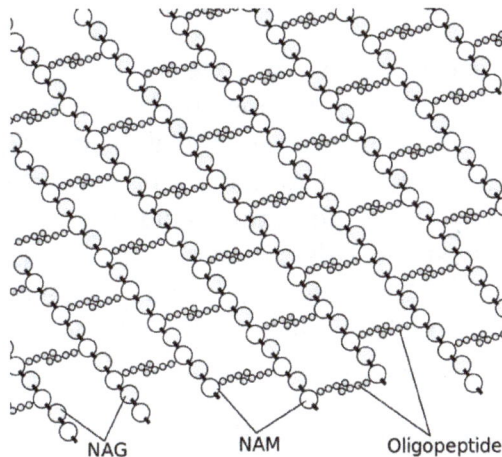

The structure of peptidoglycan, composed of N-acetylglucosamine and N-acetylmuramic acid

All gram-positive bacteria are bounded by a single-unit lipid membrane, and, in general, they contain a thick layer (20–80 nm) of peptidoglycan responsible for retaining the Gram stain. A number of other bacteria—that are bounded by a single membrane, but stain Gram-negative due to either lack of the peptidoglycan layer, as in the

Mycoplasmas, or their inability to retain the Gram stain because of their cell wall composition—also show close relationship to the Gram-positive bacteria. For the bacterial cells bounded by a single cell membrane, the term "monoderm bacteria" or "monoderm prokaryotes" has been proposed.

In contrast to gram-positive bacteria, all archetypical gram-negative bacteria are bounded by a cytoplasmic membrane and an outer cell membrane; they contain only a thin layer of peptidoglycan (2–3 nm) between these membranes. The presence of inner and outer cell membranes defines a new compartment in these cells: the periplasmic space or the periplasmic compartment. These bacteria have been designated as "diderm bacteria." The distinction between the monoderm and diderm bacteria is supported by conserved signature indels in a number of important proteins (viz. DnaK, GroEL). Of these two structurally distinct groups of bacteria, monoderms are indicated to be ancestral. Based upon a number of observations including that the gram-positive bacteria are the major producers of antibiotics and that, in general, gram-negative bacteria are resistant to them, it has been proposed that the outer cell membrane in gram-negative bacteria (diderms) has evolved as a protective mechanism against antibiotic selection pressure. Some bacteria, such as *Deinococcus*, which stain gram-positive due to the presence of a thick peptidoglycan layer and also possess an outer cell membrane are suggested as intermediates in the transition between monoderm (gram-positive) and diderm (gram-negative) bacteria. The diderm bacteria can also be further differentiated between simple diderms lacking lipopolysaccharide, the archetypical diderm bacteria where the outer cell membrane contains lipopolysaccharide, and the diderm bacteria where outer cell membrane is made up of mycolic acid.

Exceptions

In general, gram-positive bacteria are monoderms and have a single lipid bilayer whereas gram-negative bacteria are diderms and have two bilayers. Some taxa lack peptidoglycan (such as the domain Archaea, the class Mollicutes, some members of the Rickettsiales, and the insect-endosymbionts of the Enterobacteriales) and are gram-variable. This, however, does not always hold true. The *Deinococcus-Thermus* bacteria have gram-positive stains, although they are structurally similar to gram-negative bacteria with two layers. The Chloroflexi have a single layer, yet (with some exceptions) stain negative. Two related phyla to the Chloroflexi, the TM7 clade and the Ktedonobacteria, are also monoderms.

Some Firmicute species are not gram-positive. These belong to the class Mollicutes (alternatively considered a class of the phylum Tenericutes), which lack peptidoglycan (gram-indeterminate), and the class Negativicutes, which includes Selenomonas and stain gram-negative. Additionally, a number of bacterial taxa (viz. Negativicutes, Fusobacteria, Synergistetes, and Elusimicrobia) that are either part of the phylum Firmicutes or branch in its proximity are found to possess a diderm cell structure. However, a conserved signature indel (CSI) in the HSP60 (GroEL) protein distinguishes all traditional phyla of gram-negative bacteria (e.g., Proteo-bacteria, Aquificae, Chlamydiae, Bacteroidetes, Chlorobi, Cyanobacteria, Fibro-bacteres, Verrucomicrobia, Planctomycetes, Spirochetes, Acidobacteria, etc.) from these other atypical diderm bacteria, as well as other phyla of monoderm bacteria (e.g., Actinobacteria, Firmicutes, Thermotogae, Chloroflexi, etc.). The presence of this CSI in all sequenced species of conventional LPS (lipopolysaccharide)-contain-ing gram-negative bacterial phyla provides evidence that these phyla of bacteria form a monophyletic clade and that no loss of the outer membrane from any species from this group has occurred.

Pathogenesis

In the classical sense, six gram-positive genera are typically pathogenic in humans. Two of these, *Streptococcus* and *Staphylococcus*, are cocci (sphere-shaped). The remaining organisms are bacilli (rod-shaped) and can be subdivided based on their ability to form spores. The non-spore formers are *Corynebacterium* and *Listeria* (a coccobacillus), whereas *Bacillus* and *Clostridium* produce spores. The spore-forming bacteria can again be divided based on their respiration: *Bacillus* is a facultative anaerobe, while *Clostridium* is an obligate anaerobe. Also, *Rathybacter*, *Leifsonia*, and *Clavibacter* are three gram-positive genera that cause plant disease. Gram-positive bacteria are capa-ble of causing serious and sometimes fatal infections in newborn infants.

Colonies of a gram-positive pathogen of the oral cavity, *Actinomyces* sp.

Bacterial Transformation

Transformation is one of three processes for horizontal gene transfer, in which exog-enous genetic material passes from bacterium to another, the other two being conju-gation (transfer of genetic material between two bacterial cells in direct contact) and transduction (injection of foreign DNA by a bacteriophage virus into the host bacteri-um). In transformation, the genetic material passes through the intervening medium, and uptake is completely dependent on the recipient bacterium.

As of 2014 about 80 species of bacteria were known to be capable of transformation, about evenly divided between Gram-positive and Gram-negative bacteria; the number might be an overestimate since several of the reports are supported by single papers. Transformation among gram-positive bacteria has been studied in medically important species such as *Streptococcus pneumoniae*, *Streptococcus mutans*, *Staphylococcus aureus* and *Streptococcus sanguinis* and in gram-positive soil bacterium *Bacillus subtilis*.

Orthographic Note

The adjectives 'gram-positive' and 'gram-negative' derive from the surname of Hans Christian Gram; as eponymous adjectives, their initial letter can be either lowercase 'g' or capital 'G', depending on whether or not a style guide (for example that of the CDC) governs the document being written. This is further explained at *Gram staining § Orthographic note*.

Types of Gram-positive Bacteria

Lactobacillus

Lactobacillus is a genus of Gram-positive, facultative anaerobic or microaerophilic, rod-shaped, non-spore-forming bacteria. They are a major part of the lactic acid bacteria group. In humans, they constitute a significant component of the microbiota at a number of body sites. In women, *Lactobacillus* species are normally a part of the vaginal microbiota.

Biology and Biochemistry

Lactobacillus is group of rod-shaped, Gram-positive (it retains crystal violet dye), non-spore-forming, facultative anaerobe (it can produce energy through glycolysis and fermentation when oxygen is not present) bacteria. *Lactobacillus* is a member of the lactic acid bacteria group (its members convert lactose and other sugars to lactic acid).

Metabolism

Many lactobacilli operate using homofermentative metabolism (they produce only lactic acid from sugars), and some species use heterofermentative metabolism (they can produce either alcohol or lactic acid from sugars). They are aerotolerant despite the complete absence of a respiratory chain. This aerotolerance is Manganese-dependent and has been explored (and explained) in *Lactobacillus plantarum*. Many species of this genus do not require iron for growth and have an extremely high hydrogen peroxide tolerance.

Genome

Many species in this genus have had their genomes sequenced.*Lactobacillus* consists of a wealth of compound microsatellites in the coding region of the genome, which are imperfect and have variant motifs.

Taxonomy

The genus *Lactobacillus* currently contains over 180 species and encompasses a wide variety of organisms. The genus is polyphyletic, with the genus *Pediococcus* dividing the *L. casei* group, and the species *L. acidophilus*, *L. salivarius*, and *L. reuteri* being representatives of three distinct subclades. The genus *Paralactobacillus* falls within the *L. salivarius* group. In recent years, other members of the genus *Lactobacillus* (formerly known as the *Leuconostoc* branch of *Lactobacillus*) have been reclassified into the genera *Atopobium*, *Carnobacterium*, *Weissella*, *Oenococcus*, and *Leuconostoc*. More recently, the *Pediococcus* species *P. dextrinicus* has been reclassified as a *Lactobacillus* species. According to metabolism, *Lactobacillus* species can be divided into three groups:

- Obligately homofermentative (group I) including:

    o *L. acidophilus, L. delbrueckii, L. helveticus, L. salivarius*

- Facultatively heterofermentative (group II) including:

    o *L. casei, L. curvatus, L. plantarum, L. sakei*

- Obligately heterofermentative (group III) including:

    o *L. brevis, L. buchneri, L. fermentum, L. reuteri*

Clinical Uses

Lactobacillus species produce hydrogen peroxide which inhibits the growth and virulence of the fungal pathogen *Candida albicans in vitro* and *in vivo*. Following antibiotic therapy, certain *Candida* species can suppress the regrowth of *Lactobacillus* species at body sites where they cohabitate, such as in the gastrointestinal tract.

Lactobacillus species administered as a single probiotic agent is of no benefit in people with irritable bowel syndrome or Crohn's disease. When it is administered in combination with other probiotics, may help people with irritable bowel syndrome, although in a minority of cases may cause negative side effects, uncertainty remains around which type of probiotic works best, and around the size of the effect. Lactobacillus and bifidobacteria probiotics can reduce clinical symptoms of pouchitis and cholangitis. *L. acidophilus* is used to prevent necrotizing entercolitis and other neonatal infections.

Some *Lactobacillus* species have been associated with cases of dental caries. Lactic acid can corrode teeth, and the *Lactobacillus* count in saliva has been used as a "caries test" for many years. Lactobacilli characteristically cause existing carious lesions to progress, especially those in coronal caries. The issue is, however, complex, as recent studies show probiotics can allow beneficial lactobacilli to populate sites on teeth, preventing streptococcal pathogens from taking hold and inducing dental decay. The scientific research of lactobacilli in relation to oral health is a new field and only a few studies and results have been published.

Research

Research continues into the role of Lactobacillus species and the possible role it has in emotional and mental health.

Food Production

Some *Lactobacillus* species are used as starter cultures in industry for controlled fermentation in the production of yogurt, cheese, sauerkraut, pickles, beer, cider, kimchi, cocoa, kefir, and other fermented foods, as well as animal feeds. The antibacterial and antifungal activity of *Lactobacillus* species rely on production of bacteriocins and low molecular weight compounds that inhibits these microorganisms.

Sourdough bread is made either spontaneously, by taking advantage of the bacteria naturally present in flour, or by using a "starter culture", which is a symbiotic culture of yeast and lactic acid bacteria growing in a water and flour medium. The bacteria metabolize sugars into lactic acid, which lowers the pH of their environment, creating a signature "sourness" associated with yogurt, sauerkraut, etc.

In many traditional pickling processes, vegetables are submerged in brine, and salt-tolerant *Lactobacillus* species feed on natural sugars found in the vegetables. The resulting mix of salt and lactic acid is a hostile environment for other microbes, such as fungi, and the vegetables are thus preserved—remaining edible for long periods.

Lactobacilli, especially *L. casei* and *L. brevis*, are some of the most common beer spoilage organisms. They are, however, essential to the production of sour beers such as Belgian lambics and American wild ales, giving the beer a distinct tart flavor.

Staphylococcus Aureus

Staphylococcus aureus is a gram-positive coccal bacterium that is a member of the Firmicutes, and is frequently found in the nose, respiratory tract, and on the skin. It is often positive for catalase and nitrate reduction and is a facultative aerobe that can grow without the need for oxygen. Although *S. aureus* is not always pathogenic, it is a common cause of skin infections such as abscesses, respiratory infections such as sinusitis, and food poisoning. Pathogenic strains often promote infections by producing potent protein toxins, and expressing cell-surface proteins that bind and inactivate antibodies. The emergence of antibiotic-resistant strains of *S. aureus* such as methicillin-resistant *S. aureus* (MRSA) is a worldwide problem in clinical medicine.

Staphylococcus was first identified in 1880 in Aberdeen, Scotland, by the surgeon Sir Alexander Ogston in pus from a surgical abscess in a knee joint. This name was later amended to *Staphylococcus aureus* by Friedrich Julius Rosenbach, who was credited by the official system of nomenclature at the time. An estimated 20% of the human population are long-term carriers of *S. aureus* which can be found as part of the normal skin flora and in the nostrils. *S. aureus* is a normal inhabitant of the healthy lower reproductive tract of women. *S. aureus* can cause a range of illnesses, from minor skin infections, such as pimples, impetigo, boils, cellulitis, folliculitis, carbuncles, scalded skin syndrome, and abscesses, to life-threatening diseases such as pneumonia, meningitis, osteomyelitis, endocarditis, toxic shock syndrome, bacteremia, and sepsis. It is still one of the five most common causes of hospital-acquired infections and is often the cause of postsurgical wound infections. Each year, around 500,000 patients in hospitals of the United States contract a staphylococcal infection, chiefly by *S. aureus*.

Microbiology

S. aureus, ("grape-cluster ber-ry", Latin *aureus*, "golden") is a facultative anaerobic, gram-positive coccal bacteri-um also known as "golden staph" and Oro staphira. *S. aureus* is non-motile and does not form spores. In medical literature, the bacterium is often referred to as *S. aureus, Staph aureus* or *Staph A.. Staphylococcus* should not be confused with the similarly named and medically relevant genus *Streptococcus. S. aureus* appears as staphylo-cocci (grape-like clusters) when viewed through a microscope, and has large, round, golden-yellow colonies, often with hemolysis, when grown on blood agar plates. *S. aureus* reproduces asexually by binary fission. Complete separation of the daughter cells is mediated by S. aureus autolysin, and in its absence or targeted inhibition, the daughter cells remain attached to one another and appear as clusters.(Varrone JJ, de Mesy Bentley KL, Bello-Irizarry SN, Nishitani K, Mack S, Hunter JG, Kates SL, Daiss JL, Schwarz EM. Passive immunization with anti-glucosaminidase mono-clonal antibodies protects mice from implant-associated osteomyelitis by mediating opsonophagocytosis of Staphylococcus aureus megaclusters. J Orthop Res 2014;32-10:1389-96.)

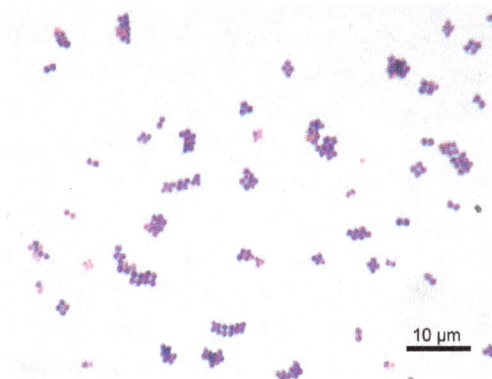

Gram stain of *S. saprophyticus* cells which typically occur in clusters:
The cell wall readily absorbs the crystal violet stain.

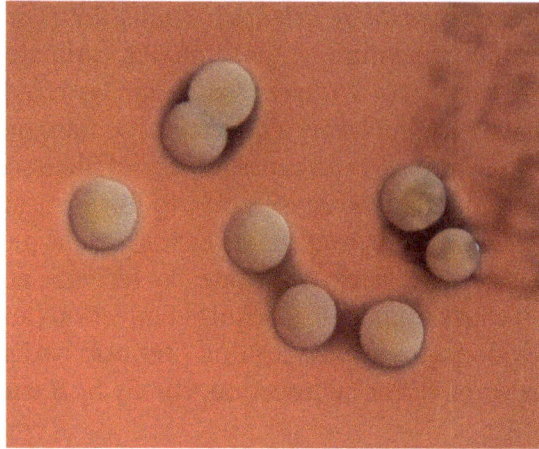

Yellow colonies of *S. aureus* on a blood agar plate, note regions of clearing around colonies caused by lysis of red cells in the agar (beta hemolysis)

S. aureus is catalase-positive (meaning it can produce the enzyme catalase). Catalase converts hydrogen peroxide (H_2O_2) to water and oxygen. Catalase-activity tests are sometimes used to distinguish staphylococci from enterococci and streptococci. Previously, *S. aureus* was differentiated from other staphylococci by the coagulase test. However, not all *S. aureus* strains are coagulase-positive and incorrect species identification can impact effective treatment and control measures.

Role in Disease

While *S. aureus* usually acts as a commensal bacterium, asymptomatically colonizing about 30% of the human population, it can sometimes cause disease. In particular, *S. aureus* is one of the most common causes of bacteremia and infective endocarditis. Additionally, it can cause various skin and soft tissue infections, particularly when skin or mucosal barriers have been breached.

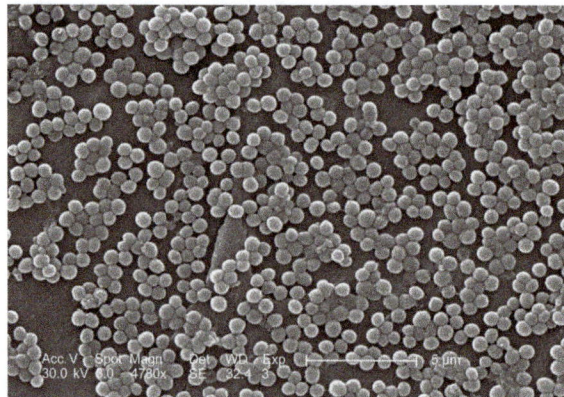

This 2005 scanning electron micrograph (SEM) depicts numerous clumps of methicillin-resistant *S. aureus* (MRSA) bacteria.

S. aureus infections can spread through contact with pus from an infected wound, skin-to-skin contact with an infected person, and contact with objects used by an infected person such as towels, sheets, clothing, or athletic equipment. Prosthetic joints put a person at particular risk of septic arthritis, staphylococcal endocarditis (infection of the heart valves), and pneumonia.

Deeply penetrating *S. aureus* infections can be severe. Strains of *S. aureus* can host phages, such as Φ-PVL (produces Panton-Valentine leukocidin), that increase virulence.

Skin Infections

Skin infections are the most common form of *S. aureus* infection. This can manifest in various ways, including small benign boils, folliculitis, impetigo, cellulitis, and more severe, invasive soft-tissue infections.

S. aureus is extremely prevalent in persons with atopic dermatitis. It is mostly found in fertile, active places, including the armpits, hair, and scalp. Large pimples that appear in those areas may exacerbate the infection if lacerated. This can lead to staphylococcal scalded skin syndrome. A severe form of this, Ritter's disease, can be observed in neonates.

The presence of *S. aureus* in persons with atopic dermatitis is not an indication to treat with oral antibiotics, as evidence has not shown this to give benefit to the patient. The relationship between *S. aureus* and atopic dermatitis is unclear.

Food Poisoning

S. aureus is also responsible for food poisoning. It is capable of generating toxins that produce food poisoning in the human body.

Bone and Joint Infections

S. aureus is the bacterium that is commonly responsible for all major bone and joint infections. This manifests in one of three forms: osteomyelitis, septic arthritis and prosthetic joint infection.

Bacteremia

S. aureus is a leading cause of bloodstream infections throughout much of the industrialized world. Infection is generally associated with breakages in the skin or mucosal membranes due to surgery, injury, or use of intravascular devices such as catheters, hemodialysis machines, or injected drugs. Once the bacteria have entered the bloodstream, they can infect various organs, causing infective endocarditis, septic arthritis, and osteomyelitis. This disease is particularly prevalent and severe in the very young and very old.

Without antibiotic treatment, *S. aureus* bacteremia has a case fatality rate around 80%. With antibiotic treatment, case fatality rates range from 15% to 50% depending on the age and health of the patient, as well as the antibiotic resistance of the *S. aureus* strain.

Animal Infections

S. aureus can survive on dogs, cats, and horses, and can cause bumblefoot in chickens. Some believe health-care workers' dogs should be considered a significant source of antibiotic-resistant *S. aureus*, especially in times of outbreak.

S. aureus is one of the causal agents of mastitis in dairy cows. Its large polysaccharide capsule protects the organism from recognition by the cow's immune defenses.

Virulence Factors

Enzymes

S. aureus produces various enzymes such as coagulase (bound and free coagulases) which clots plasma and coats the bacterial cell, probably to prevent phagocytosis. Hyaluronidase (also known as spreading factor) breaks down hyaluronic acid and helps in spreading it. *S. aureus* also produces deoxyribonuclease, which breaks down the DNA, lipase to digest lipids, staphylokinase to dissolve fibrin and aid in spread, and beta-lactamase for drug resistance.

Toxins

Depending on the strain, *S. aureus* is capable of secreting several exotoxins, which can be categorized into three groups. Many of these toxins are associated with specific diseases.

Superantigens

> (PTSAgs) have superantigen activities that induce toxic shock syndrome (TSS). This group includes the toxin TSST-1, enterotoxin type B, which causes TSS associated with tampon use. This is characterized by fever, erythematous rash, hypotension, shock, multiple organ failure, and skin desquamation. Lack of antibody to TSST-1 plays a part in the pathogenesis of TSS. Other strains of *S. aureus* can produce an enterotoxin that is the causative agent of *S. aureus* gastroenteritis. This gastroenteritis is self-limiting, characterized by vomiting and diarrhea one to six hours after ingestion of the toxin, with recovery in eight to 24 hours. Symptoms include nausea, vomiting, diarrhea, and major abdominal pain.

Exfoliative toxins

> EF toxins are implicated in the disease staphylococcal scalded-skin syndrome (SSSS), which occurs most commonly in infants and young children. It also may

occur as epidemics in hospital nurseries. The protease activity of the exfoliative toxins causes peeling of the skin observed with SSSS.

Other toxins

Staphylococcal toxins that act on cell membranes include alpha toxin, beta toxin, delta toxin, and several bicomponent toxins. The bicomponent toxin Panton-Valentine leukocidin (PVL) is associated with severe necrotizing pneumonia in children. The genes encoding the components of PVL are encoded on a bacteriophage found in community-associated MRSA strains.

Other Immunoevasive Strategies

Protein A

Protein A is anchored to staphylococcal peptidoglycan pentaglycine bridges (chains of five glycine residues) by the transpeptidase sortase A. Protein A, an IgG-binding protein, binds to the Fc region of an antibody. In fact, studies involving mutation of genes coding for protein A resulted in a lowered virulence of *S. aureus* as measured by survival in blood, which has led to speculation that protein A-contributed virulence requires binding of antibody Fc regions.

Protein A in various recombinant forms has been used for decades to bind and purify a wide range of antibodies by immunoaffinity chromatography. Transpeptidases, such as the sortases responsible for anchoring factors like protein A to the staphylococcal peptidoglycan, are being studied in hopes of developing new antibiotics to target MRSA infections.

S. aureus on trypticase soy agar: The strain is producing a yellow pigment staphyloxanthin.

Staphylococcal pigments

Some strains of *S. aureus* are capable of producing staphyloxanthin — a golden-coloured carotenoid pigment. This pigment acts as a virulence factor, primarily by being a bacterial antioxidant which helps the microbe evade the reactive oxygen species which the host immune system uses to kill pathogens.

Mutant strains of *S. aureus* modified to lack staphyloxanthin are less likely to survive incubation with an oxidizing chemical, such as hydrogen peroxide, than pigmented strains. Mutant colonies are quickly killed when exposed to human neutrophils, while many of the pigmented colonies survive. In mice, the pigmented strains cause lingering abscesses when inoculated into wounds, whereas wounds infected with the unpigmented strains quickly heal.

These tests suggest the *Staphylococcus* strains use staphyloxanthin as a defence against the normal human immune system. Drugs designed to inhibit the production of staphyloxanthin may weaken the bacterium and renew its susceptibility to antibiotics. In fact, because of similarities in the pathways for biosynthesis of staphyloxanthin and human cholesterol, a drug developed in the context of cholesterol-lowering therapy was shown to block *S. aureus* pigmentation and disease progression in a mouse infection model.

Classical Diagnosis

Depending upon the type of infection present, an appropriate specimen is obtained accordingly and sent to the laboratory for definitive identification by using biochemical or enzyme-based tests. A Gram stain is first performed to guide the way, which should show typical gram-positive bacteria, cocci, in clusters. Second, the isolate is cultured on mannitol salt agar, which is a selective medium with 7–9% NaCl that allows *S. aureus* to grow, producing yellow-colored colonies as a result of mannitol fermentation and subsequent drop in the medium's pH.

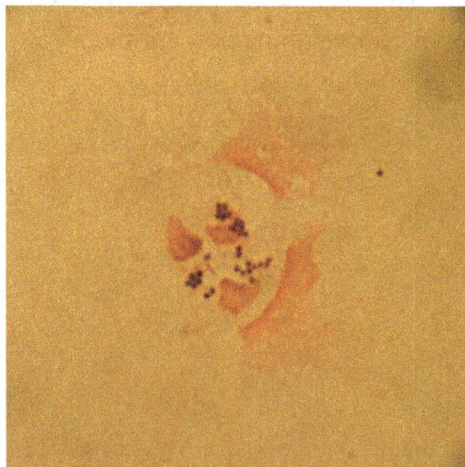

Typical gram-positive cocci, in clusters, from a sputum sample, Gram stain

Furthermore, for differentiation on the species level, catalase (positive for all *Staphylococcus* species), coagulase (fibrin clot formation, positive for *S. aureus*), DNAse (zone of clearance on DNAse agar), lipase (a yellow color and rancid odor smell), and phosphatase (a pink color) tests are all done. For staphylococcal food poisoning, phage typing can be performed to determine whether the staphylococci recovered from the food were the source of infection.

Rapid Diagnosis and Typing

Diagnostic microbiology laboratories and reference laboratories are key for identifying outbreaks and new strains of *S. aureus*. Recent genetic advances have enabled reliable and rapid techniques for the identification and characterization of clinical isolates of *S. aureus* in real time. These tools support infection control strategies to limit bacterial spread and ensure the appropriate use of antibiotics. Quantitative PCR is increasingly being used to identify outbreaks of infection.

When observing the evolvement of *S. aureus* and its ability to adapt to each modified antibiotic, two basic methods known as "band-based" or "sequence-based" are employed. Keeping these two methods in mind, other methods such as multilocus sequence typing (MLST), pulsed-field gel electrophoresis (PFGE), bacteriophage typing, spa locus typing, and SCCmec typing are often conducted more than others. With these methods, it can be determined where strains of MRSA originated and also where they are currently.

With MLST, this technique of typing uses fragments of several housekeeping genes known as *aroE, glpF, gmk, pta, tip,* and *yqiL*. These sequences are then assigned a number which give to a string of several numbers that serve as the allelic profile. Although this is a common method, a limitation about this method is the maintenance of the microarray which detects newly allelic profiles, making it a costly and time-consuming experiment.

With PFGE, a method which is still very much used dating back to its first success in 1980s, remains capable of helping differentiate MRSA isolates. To accomplish this, the technique uses multiple gel electrophoresis, along with a voltage gradient to display clear resolutions of molecules. The *S. aureus* fragments then transition down the gel, producing specific band patters that are later compared with other isolates in hopes of identifying related strains. Limitations of the method include practical difficulties with uniform band patterns and PFGE sensitivity as a whole.

Spa locus typing is also considered a popular technique that uses a single locus zone in a polymorphic region of *S. aureus* to distinguish any form of mutations. Although this technique is often inexpensive and less time-consuming, the chance of losing discriminatory power makes it hard to differentiate between MLST CCs exemplifies a crucial limitation.

Treatment and Antibiotic Resistance

The treatment of choice for *S. aureus* infection is penicillin. An antibiotic derived from *Penicillum* fungus, penicillin inhibits the formation of peptidoglycan cross-linkages that provide the rigidity and strength in a bacterial cell wall. The four-membered β-lactam ring of penicillin is bound to enzyme DD-transpeptidase, an enzyme that when functional, cross-links chains of peptidoglycan that form bacterial cell walls. The binding of β-lactam to DD-transpeptidase inhibits the enzyme's functionality and it can no longer catalyze the formation of the cross-links. As a result, cell wall formation and degradation are imbalanced, thus resulting in cell death. In most countries, however, penicillin resistance is extremely common, and first-line therapy is most commonly a penicillinase-resistant β-lactam antibiotic (for example, oxacillin or flucloxacillin, both of which have the same mechanism of action as penicillin). Combination therapy with gentamicin may be used to treat serious infections, such as endocarditis, but its use is controversial because of the high risk of damage to the kidneys. Honey and propolis produced by the South American bee *Tetragonisca angustula* has also been found to have antibacterial activity towards S. aureus. The duration of treatment depends on the site of infection and on severity.

Antibiotic resistance in *S. aureus* was uncommon when penicillin was first introduced in 1943. Indeed, the original Petri dish on which Alexander Fleming of Imperial College London observed the antibacterial activity of the *Penicillium* fungus was growing a culture of *S. aureus*. By 1950, 40% of hospital *S. aureus* isolates were penicillin-resistant; by 1960, this had risen to 80%.

MRSA and often pronounced, is one of a number of greatly feared strains of *S. aureus* which have become resistant to most β-lactam antibiotics. For this reason, vancomycin, a glycopeptide antibiotic, is commonly used to combat MRSA. Vancomycin inhibits the synthesis of peptidoglycan, but unlike β-lactam antibiotics, glycopeptide antibiotics target and bind to amino acids in the cell wall, preventing peptidoglycan cross-linkages from forming. MRSA strains are most often found associated with institutions such as hospitals, but are becoming increasingly prevalent in community-acquired infections. A recent study by the Translational Genomics Research Institute showed that nearly half (47%) of the meat and poultry in U.S. grocery stores were contaminated with *S. aureus*, with more than half (52%) of those bacteria resistant to antibiotics. This resistance is commonly caused by the widespread use of antibiotics in the husbandry of livestock, including prevention or treatment of an infection, as well as promoting growth.

Researchers from ETH Zurich have created the endolysin Staphefekt SA.100, which is active against *S. aureus*, including MRSA.

Minor skin infections can be treated with triple antibiotic ointment.

Mechanisms of Antibiotic Resistance

Staphylococcal resistance to penicillin is mediated by penicillinase (a form of β-lactamase) production: an enzyme that cleaves the β-lactam ring of the penicillin molecule, rendering the antibiotic ineffective. Penicillinase-resistant β-lactam antibiotics, such as methicillin, nafcillin, oxacillin, cloxacillin, dicloxacillin, and flucloxacillin, are able to resist degradation by staphylococcal penicillinase.

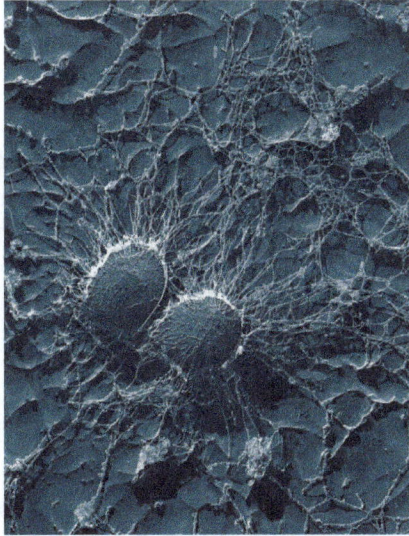

Bacterial cells of *S. aureus*, which is one of the causal agents of mastitis in dairy cows: Its large capsule protects the organism from attack by the cow's immunological defenses.

Resistance to methicillin is mediated via the *mec* operon, part of the staphylococcal cassette chromosome mec (SCC*mec*). Resistance is conferred by the *mecA* gene, which codes for an altered penicillin-binding protein (PBP2a or PBP2') that has a lower affinity for binding β-lactams (penicillins, cephalosporins, and carbapenems). This allows for resistance to all β-lactam antibiotics, and obviates their clinical use during MRSA infections. As such, the glycopeptide vancomycin is often deployed against MRSA.

Aminoglycoside antibiotics, such as kanamycin, gentamicin, streptomycin, etc., were once effective against staphylococcal infections until strains evolved mechanisms to inhibit the aminoglycosides' action, which occurs via protonated amine and/or hydroxyl interactions with the ribosomal RNA of the bacterial 30S ribosomal subunit. Three main mechanisms of aminoglycoside resistance mechanisms are currently and widely accepted: aminoglycoside modifying enzymes, ribosomal mutations, and active efflux of the drug out of the bacteria.

Aminoglycoside-modifying enzymes inactivate the aminoglycoside by covalently attaching either a phosphate, nucleotide, or acetyl moiety to either the amine or the alcohol key functional group (or both groups) of the antibiotic. This changes the charge or sterically hinders the antibiotic, decreasing its ribosomal binding affinity. In *S. aureus*, the best-characterized aminoglycoside-modifying enzyme is aminoglycoside adenylyltransferase 4' IA (*ANT(4')IA*). This enzyme has been solved by X-ray crystallography. The enzyme is able to attach an adenyl moiety to the 4' hydroxyl group of many aminoglycosides, including kamamycin and gentamicin.

Glycopeptide resistance is mediated by acquisition of the *vanA* gene, which originates from the enterococci and codes for an enzyme that produces an alternative peptidoglycan to which vancomycin will not bind.

Today, *S. aureus* has become resistant to many commonly used antibiotics. In the UK, only 2% of all *S. aureus* isolates are sensitive to penicillin, with a similar picture in the rest of the world. The β-lactamase-resistant penicillins (methicillin, oxacillin, cloxacillin, and flucloxacillin) were developed to treat penicillin-resistant *S. aureus*, and are still used as first-line treatment. Methicillin was the first antibiotic in this class to be used (it was introduced in 1959), but, only two years later, the first case of MRSA was reported in England.

Despite this, MRSA generally remained an uncommon finding, even in hospital settings, until the 1990s, when the MRSA prevalence in hospitals exploded, and it is now endemic.

MRSA infections in both the hospital and community setting are commonly treated with non-β-lactam antibiotics, such as clindamycin (a lincosamine) and co-trimoxazole (also commonly known as trimethoprim/sulfamethoxazole). Resistance to these antibiotics has also led to the use of new, broad-spectrum anti-gram-positive antibiotics, such as linezolid, because of its availability as an oral drug. First-line treatment for serious invasive infections due to MRSA is currently glycopeptide antibiotics (vancomycin and teicoplanin). A number of problems with these antibiotics occur, such as the need for intravenous administration (no oral preparation is available), toxicity, and the need to monitor drug levels regularly by blood tests. Also, glycopeptide antibiotics do not penetrate very well into infected tissues (this is a particular concern with infections of the brain and meninges and in endocarditis). Glycopeptides must not be used to treat methicillin-sensitive *S. aureus* (MSSA), as outcomes are inferior.

Because of the high level of resistance to penicillins and because of the potential for MRSA to develop resistance to vancomycin, the U.S. Centers for Disease Control and Prevention has published guidelines for the appropriate use of vancomycin. In situations where the incidence of MRSA infections is known to be high, the attending physician may choose to use a glycopeptide antibiotic until the identity of the infecting organism is known. After the infection is confirmed to be due to a methicillin-susceptible strain of *S. aureus*, treatment can be changed to flucloxacillin or even penicillin], as appropriate.

Vancomycin-resistant *S. aureus* (VRSA) is a strain of *S. aureus* that has become resistant to the glycopeptides. The first case of vancomycin-intermediate *S. aureus* (VISA) was reported in Japan in 1996; but the first case of *S. aureus* truly resistant to glycopeptide antibiotics was only reported in 2002. Three cases of VRSA infection had been reported in the United States as of 2005.

In a recent study done on mice, a group of scientists saw that polyunsaturated fatty acids helped survival rates among mice and overall keep bacteria count lower when compared to other mice when going through sepsis. they did this experiment in efforts to help find other ways to combat the amount of growing antibiotic resistant strains.

Small non-coding RNA SprX was shown to influence *s. aureus* antibiotic resistance to Vancomycin and Teicoplanin.

Carriage of S. Aureus

About one-third of the U.S. population are carriers of *S. aureus.*

The carriage of *S. aureus* is an important source of hospital-acquired infection (also called nosocomial) and community-acquired MRSA. Although *S. aureus* can be present on the skin of the host, a large proportion of its carriage is through the anterior nares of the nasal passages and can further be present in the ears. The ability of the nasal passages to harbour *S. aureus* results from a combination of a weakened or defective host immunity and the bacterium's ability to evade host innate immunity. Nasal carriage is also implicated in the occurrence of staph infections.

Infection Control

Spread of *S. aureus* (including MRSA) generally is through human-to-human contact, although recently some veterinarians have discovered the infection can be spread through pets, with environmental contamination thought to play a relatively unimportant part. Emphasis on basic hand washing techniques are, therefore, effective in preventing its transmission. The use of disposable aprons and gloves by staff reduces skin-to-skin contact, so further reduces the risk of transmission.

Recently, myriad cases of *S. aureus* have been reported in hospitals across America. Transmission of the pathogen is facilitated in medical settings where healthcare worker hygiene is insufficient. *S. aureus* is an incredibly hardy bacterium, as was shown in a study where it survived on polyester for just under three months; polyester is the main material used in hospital privacy curtains.

The bacteria are transported on the hands of healthcare workers, who may pick them up from a seemingly healthy patient carrying a benign or commensal strain of *S. aureus*, and then pass it on to the next patient being treated. Introduction of the bacteria into the bloodstream can lead to various complications, including endocarditis, meningitis, and, if it is widespread, sepsis.

Ethanol has proven to be an effective topical sanitizer against MRSA. Quaternary ammonium can be used in conjunction with ethanol to increase the duration of the sanitizing action. The prevention of nosocomial infections involves routine and terminal cleaning. Nonflammable alcohol vapor in CO_2 NAV-CO_2 systems have an advantage, as they do not attack metals or plastics used in medical environments, and do not contribute to antibacterial resistance.

An important and previously unrecognized means of community-associated MRSA colonization and transmission is during sexual contact.

S. aureus is killed in one minute at 78 °C and in ten minutes at 64 °C.

Top common bacterium in each industry
Catering industry
Vibrio parahaemolyticus , *S. aureus*, *Bacillus cereus*
Medical industry
Escherichia coli, *S. aureus* , *Pseudomonas aeruginosa*

Natural Genetic Transformation

Natural genetic transformation is a sexual process involving DNA transfer from one bacterium to another through the intervening medium, and the integration of the donor sequence into the recipient genome by homologous recombination. *S. aureus* was found to be capable of natural genetic transformation, but only at low frequency under the experimental conditions employed. Further studies suggested that the development of competence for natural genetic transformation may be substantially higher under appropriate conditions, yet to be discovered.

Research

As of 2015, no approved vaccine exists against *S. aureus*. Early clinical trials have been conducted for several vaccines candidates such as Nabi's StaphVax and PentaStaph, Intercell's / Merck's V710, VRi's SA75, and others.

While some of these vaccines candidates have shown immune responses, other aggravated an infection by *S. aureus*. To date, none of these candidates provides protection against a *S. aureus* infection. The development of Nabi's StaphVax was stopped in 2005 after phase III trials failed. Intercell's first V710 vaccine variant was terminated during phase II/III after higher mortality and morbidity were observed among patients who developed *S. aureus* infection.

Nabi's enhanced *S. aureus* vaccines candidate PentaStaph was sold in 2011 to GlaxoSmithKline Biologicals S.A. The current status of PentaStaph is unclear.

In 2010, GlaxoSmithKline started a phase 1 blind study to evaluate its GSK2392103A vaccine. As of 2016, this vaccine is no longer under active development.

Pfizer's *S. aureus* four-antigen vaccine SA4Ag has granted fast track designation by the U.S. Food and Drug Administration in February 2014. In 2015, Pfizer has commenced a phase 2b trial regarding the SA4Ag vaccine.

Novartis Vaccines and Diagnostics, a former division of Novartis and now part of GlaxoSmithKline, published in 2015 promising pre-clinical results of their four-component Staphylococcus aureus vaccine, 4C-staph.

Bacillus

Bacillus is a genus of gram-positive, rod-shaped bacteria and a member of the phylum Firmicutes. *Bacillus* species can be obligate aerobes (oxygen reliant), or facultative anaerobes (having the ability to be aerobic or anaerobic). They will test positive for the enzyme catalase when there has been oxygen used or present. Ubiquitous in nature, *Bacillus* includes both free-living (nonparasitic) and parasitic pathogenic species. Under stressful environmental conditions, the bacteria can produce oval endospores that are not true 'spores', but to which the bacteria can reduce themselves and remain in a dormant state for very long periods. These characteristics originally defined the genus, but not all such species are closely related, and many have been moved to other genera of the Firmicutes.

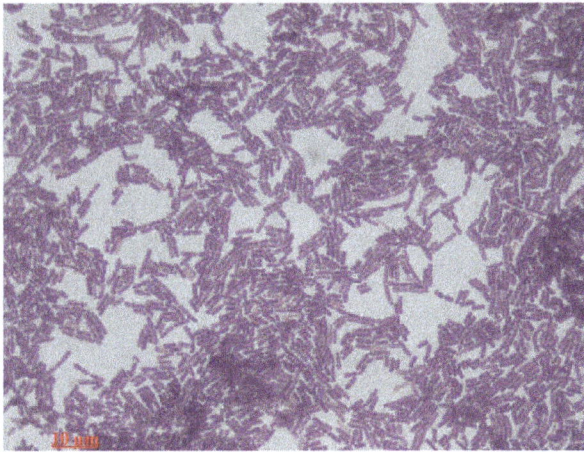

Gram stain of a *Bacillus* species

Many species of *Bacillus* can produce copious amounts of enzymes which are made use of in different industries. Some species can form intracellular inclusions of polyhydroxyalkanoates under certain adverse environmental conditions, as in a lack of elements such as phosphorus, nitrogen, or oxygen combined with an excessive supply of carbon sources.

B. subtilis has proved a valuable model for research. Other species of *Bacillus* are important pathogens, causing anthrax and food poisoning.

Industrial Significance

Many *Bacillus* species are able to secrete large quantities of enzymes. *Bacillus amyloliquefaciens* is the source of a natural antibiotic protein barnase (a ribonuclease), alpha amylase used in starch hydrolysis, the protease subtilisin used with detergents, and the BamH1 restriction enzyme used in DNA research.

A portion of the *Bacillus thuringiensis* genome was incorporated into corn (and cotton) crops. The resulting GMOs are therefore resistant to some insect pests.

Use as Model Organism

Bacillus subtilis is one of the best understood prokaryotes, in terms of molecular and cellular biology. Its superb genetic amenability and relatively large size have provided the powerful tools required to investigate a bacterium from all possible aspects. Recent improvements in fluorescent microscopy techniques have provided novel insight into the dynamic structure of a single cell organism. Research on *B. subtilis* has been at the forefront of bacterial molecular biology and cytology, and the organism is a model for differentiation, gene/protein regulation, and cell cycle events in bacteria.

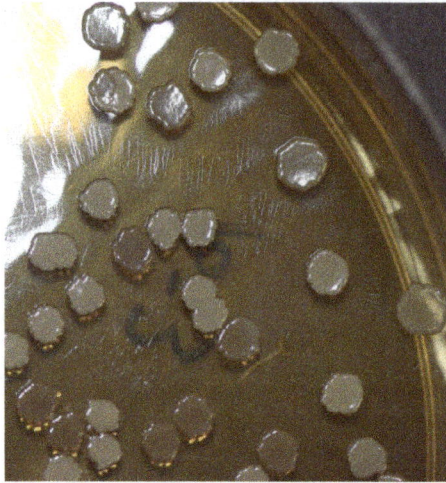

Colonies of the model species *Bacillus subtilis* on an agar plate.

Ecological Significance

Bacillus species are almost ubiquitous in nature, e.g. in soil, but also occur in extreme environments such as high pH (*B. alcalophilus*), high temperature (*B. thermophilus*), or high salt (*B. halodurans*). *B. thuringiensis* produces a toxin that can kill insects and thus has been used as insecticide.

Clinical Significance

Two *Bacillus* species are considered medically significant: *B. anthracis*, which causes anthrax, and *B. cereus*, which causes food poisoning similar to that caused by *Staphylococcus*. A third species, *B. thuringiensis*, is an important insect pathogen, and is sometimes used to control insect pests. The type species is *B. subtilis*, an important model organism. It is also a notable food spoiler, causing ropiness in bread and related food. Some environmental and commercial strains of *B. coagulans* may play a role in food spoilage of highly acidic, tomato-based products.

An easy way to isolate *Bacillus* species is by placing nonsterile soil in a test tube with water, shaking, placing in melted mannitol salt agar, and incubating at room temperature for at least a day. Colonies are usually large, spreading, and irregularly shaped.

Under the microscope, the *Bacillus* cells appear as rods, and a substantial portion of the cells usually contain oval endospores at one end, making it bulge.

Cell Wall

The cell wall of *Bacillus* is a structure on the outside of the cell that forms the second barrier between the bacterium and the environment, and at the same time maintains the rod shape and withstands the pressure generated by the cell's turgor. The cell wall is composed of teichoic and teichuronic acids. *B. subtilis* is the first bacterium for which the role of an actin-like cytoskeleton in cell shape determination and peptidoglycan synthesis was identified, and for which the entire set of peptidoglycan-synthesizing enzymes was localised. The role of the cytoskeleton in shape generation and maintenance is important

Phylogeny

The genus *Bacillus* was named in 1835 by Christian Gottfried Ehrenberg, to contain rod-shaped (bacillus) bacteria. He had seven years earlier named the genus *Bacterium*. *Bacillus* was later amended by Ferdinand Cohn to further describe them as spore-forming, Gram-positive, aerobic or facultatively anaerobic bacteria. Like other genera associated with the early history of microbiology, such as *Pseudomonas* and *Vibrio*, the 266 species of *Bacillus* are ubiquitous. The genus has a very large ribosomal 16S diversity and is environmentally diverse.

Several studies have tried to reconstruct the phylogeny of the genus. The *Bacillus*-specific study with the most diversity covered is by Xu and Cote using 16S and the ITS regions, where they divide the genus into 10 groups, which includes the nested genera *Paenibacillus*, *Brevibacillus*, *Geobacillus*, *Marinibacillus* and *Virgibacillus*. However, the tree constructed by the living tree project, a collaboration between ARB-Silva and LPSN where a 16S (and 23S if available) tree of all validated species was constructed, the genus *Bacillus* contains a very large number of nested taxa and majorly in both 16S and 23S it is paraphyletic to the Lactobacillales (*Lactobacillus*, *Streptococcus*, *Staphylococcus*, *Listeria*, etc.), due to *Bacillus coahuilensis* and others. A gene concatenation study found similar results to Xu and Cote, but with a much more limited number of species in terms of groups, but used *Listeria* as an outgroup, so in light of the ARB tree, it may be "inside-out".

One clade, formed by *B. anthracis*, *B. cereus*, *B. mycoides*, *B. pseudomycoides*, *B. thuringiensis*, and *B. weihenstephanensis* under current classification standards, should be a single species (within 97% 16S identity), but due to medical reasons, they are considered separate species, an issue also present for four species of *Shigella* and *Escherichia coli*.

Bacillus Cereus

Bacillus cereus is a Gram-positive, rod-shaped, aerobic, motile, beta hemolytic bacterium commonly found in soil and food. Some strains are harmful to humans and cause foodborne illness, while other strains can be beneficial as probiotics for animals.. It is the cause of "fried rice syndrome", as the bacteria are classically contracted from fried rice dishes that have been sitting at room temperature for hours (such as at a buffet). *B. cereus* bacteria are facultative anaerobes, and like other members of the genus *Bacillus*, can produce protective endospores. Its virulence factors include cereolysin and phospholipase C.

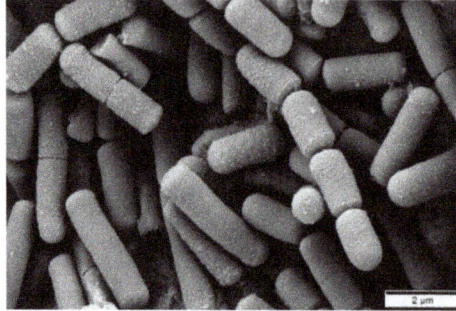

Electron micrograph of *Bacillus cereus*

It was from this species that two new enzymes, named AlkC and AlkD, which are involved in DNA repair, were discovered in 2006.

History

Colonies of *Bacillus cereus* were originally isolated from an agar plate left exposed to the air in a cow shed. In the 2010s, examination of warning letters issued by the US Food and Drug Administration issued to pharmaceutical manufacturing facilities addressing facility microbial contamination revealed that the most common contaminant was *B. cereus*.

Ecology

B. cereus competes with other microorganisms such as *Salmonella* and *Campylobacter* in the gut, so its presence reduces the numbers of those microorganisms. In food animals such as chickens, rabbits and pigs, some harmless strains of *B. cereus* are used as a probiotic feed additive to reduce *Salmonella* in the intestines and cecum. This improves the animals' growth as well as food safety for humans who eat their meat.

Bacillus cereus and other members of *Bacillus* are not easily killed by alcohol; in fact, they have been known to colonize distilled liquors and alcohol-soaked swabs and pads in numbers sufficient to cause infection.

Some strains of *B. cereus* produce cereins, bacteriocins active against different *B. cereus* strains or other Gram-positive bacteria.

Reproduction

At 30 °C (86 °F), a population of *B. cereus* can double in as little as 20 minutes or as long as 3 hours, depending on the food product.

Pathogenesis

B. cereus is responsible for a minority of foodborne illnesses (2–5%), causing severe nausea, vomiting, and diarrhea. *Bacillus* foodborne illnesses occur due to survival of the bacterial endospores when food is improperly cooked. Cooking temperatures less than or equal to 100 °C (212 °F) allow some *B. cereus* spores to survive. This problem is compounded when food is then improperly refrigerated, allowing the endospores to germinate. Cooked foods not meant for either immediate consumption or rapid cooling and refrigeration should be kept at temperatures below 10 °C (50 °F) or above 50 °C (122 °F). Germination and growth generally occur between 10 °C and 50 °C, though some strains are psychrotrophic. Bacterial growth results in production of enterotoxins, one of which is highly resistant to heat and acids (pH levels between 2 and 11); ingestion leads to two types of illness, diarrheal and emetic (vomiting) syndrome.

- The diarrheal type is associated with a wide range of foods, has an 8.0- to 16-hour incubation time, and is associated with diarrhea and gastrointestinal pain. Also known as the 'long-incubation' form of *B. cereus* food poisoning, it might be difficult to differentiate from poisoning caused by *Clostridium perfringens*. Enterotoxin can be inactivated after heating at 56 °C (133 °F) for 5 minutes however it is unclear whether its presence in food causes the symptom since it degrades in stomach enzymes; its subsequent production by surviving *B. cereus* spores within the small intestine may be the cause of illness.

- The 'emetic' form is commonly caused by rice cooked for a time and temperature insufficient to kill any spores present, then improperly refrigerated. It can produce a toxin, cereulide, which is not inactivated by later reheating. This form leads to nausea and vomiting one to five hours after consumption. It can be difficult to distinguish from other short-term bacterial foodborne intoxications such as by *Staphylococcus aureus*. Emetic toxin can withstand 121 °C (250 °F) for 90 minutes.

The diarrhetic syndromes observed in patients are thought to stem from the three toxins: hemolysin BL (Hbl), nonhemolytic enterotoxin (Nhe) and cytotoxin K (CytK). The *nhe/hbl/cytK* genes are located on the chromosome of the bacteria. Transcription of these genes is controlled by *PlcR*. These genes occur in the taxonomically related *B. thuringiensis* and *B. anthracis*, as well. These enterotoxins are all produced in the small intestine of the host, thus thwarting digestion by host endogenous enzymes. The Hbl and Nhe toxins are pore-forming toxins closely related to ClyA of *E. coli*. The proteins exhibit a conformation known as "beta-barrel" that can insert into cellular membranes due to a hydrophobic exterior, thus creating pores with hydrophilic interiors. The effect is loss of cellular membrane potential and eventually cell death. CytK is a pore-forming protein more related to other hemolysins.

The timing of the toxin production was previously thought to be possibly responsible for the two different courses of disease, but in fact the emetic syndrome is caused by a toxin, cereulide, found only in emetic strains and is not part of the "standard toolbox" of *B. cereus*. Cereulide is a cyclic polypeptide containing three repeats of four amino acids: D-oxy-Leu—D-Ala—L-oxy-Val—L-Val (similar to valinomycin produced by *Streptomyces griseus*) produced by nonribosomal peptide synthesis. Cereulide is believed to bind to 5-hydroxytryptamine 3 (5-HT3) serotonin receptors, activating them and leading to increased afferent vagus nerve stimulation. It was shown independently by two research groups to be encoded on multiple plasmids: pCERE01 or pBCE4810. Plasmid pBCE4810 shares homology with the *Bacillus anthracis* virulence plasmid pXO1, which encodes the anthrax toxin. Periodontal isolates of *B. cereus* also possess distinct pXO1-like plasmids. Like most of cyclic peptides containing nonproteogenic amino acids, cereulid is resistant to heat, proteolysis, and acid conditions.

B. cereus is also known to cause difficult-to-eradicate chronic skin infections, though less aggressive than necrotizing fasciitis. *B. cereus* can also cause keratitis.

Diagnosis

In case of foodborne illness, the diagnosis of *B. cereus* can be confirmed by the isolation of more than 10^5 *B. cereus* organisms per gram from epidemiologically implicated food, but such testing is often not done because the illness is relatively harmless and usually self-limiting.

Prognosis

Most emetic patients recover within six to 24 hours, but in some cases, the toxin can be fatal. In 2014, 23 neonates receiving total parenteral nutrition contaminated with *B. cereus* developed septicaemia, with three of the infants later dying as a result of infection.

Corynebacterium

Corynebacterium is a genus of bacteria that are gram-positive and aerobic. They are bacilli (rod-shaped), and in some phases of life they are, more particularly, club-shaped, which inspired the genus name (*coryneform* means "club-shaped").

They are widely distributed in nature in the microbiota of animals (including the human microbiota) and are mostly innocuous. Some are useful in industrial settings such as *C. glutamicum*. Others can cause human disease, including most notably diphtheria, which is caused by *C. diphtheriae*. As with various species of a microbiota (including their cousins in the genera *Arcanobacterium* and *Trueperella*), they usually are not pathogenic but can occasionally opportunistically capitalize on atypical access to tissues (via wounds) or weakened host defenses.

Taxonomy

The genus *Corynebacterium* was created by Lehmann and Neumann in 1896 as a taxonomic group to contain the bacterial rods responsible for causing diphtheria. The genus was defined based on morphological characteristics. Based on studies of 16S-rRNA, they have been grouped into the subdivision of gram-positive eubacteria with high G:C content, with close phylogenetic relationship to *Arthrobacter, Mycobacterium, Nocardia*, and *Streptomyces*.

The term comes from the *corönë* ("knotted rod") and *bacterion* ("rod"). The term "diphtheroids" is used to represent corynebacteria that are nonpathogenic; for example, *C. diphtheriae* would be excluded (reference?). The term diphtheroid comes from *diphthera*—prepared hide, leather.

Genomics

Comparative analysis of corynebacterial genomes has led to the identification of several conserved signature indels which are unique to the genus. Two examples of these conserved signature indels are a two-amino-acid insertion in a conserved region of the enzyme phosphoribose diphosphate:decaprenyl-phosphate phosphoribosyltransferase and a three-amino-acid insertion in acetate kinase, both of which are found only in *Corynebacterium* species. Both of these indels serve as molecular markers for species of the genus *Corynebacterium*. Additionally, 16 conserved signature proteins, which are uniquely found in *Corynebacterium* species, have been identified. Three of the conserved signature proteins have homologs found in the *Dietzia* genus, which is believed to be the closest related genus to *Corynebacterium*. In phylogenetic trees based on concatenated protein sequences or 16S rRNA, the genus *Corynebacterium* forms a distinct clade, within which is a distinct subclade, cluster I. The cluster is made up of the species *C. diptheriae, C. pseudotuberculosis, C. ulcerans, C. aurimucosum, C. glutamicum,* and *C. efficiens.* This cluster is distinguished by several conserved signature indels, such as a two-amino-acid insertion in LepA and a seven- or eight-amino-acid insertions in RpoC. Also, 21 conserved signature proteins are found only in members of cluster I. Another cluster has been proposed, consisting of *C. jeikeium* and *C. urealyticum*, which is supported by the presence of 19 distinct conserved signature proteins which are unique to these two species.

Characteristics

The principal features of the *Corynebacterium* genus were described by Collins and Cummins in 1986. They are gram-positive, catalase-positive, nonspore-forming, nonmotile, rod-shaped bacteria that are straight or slightly curved. Metachromatic granules are usually present representing stored phosphate regions. Their size falls between 2 and 6 µms in length and 0.5 µm in diameter. The bacteria group together in a characteristic way, which has been described as the form of a "V", "palisades", or "Chinese letters". They may also appear elliptical. They are aerobic or facultatively anaerobic, chemoorganotrophs, with a 51–65% genomic G:C content. They are pleomorphic through their lifecycles, they occur in various lengths, and they frequently have thickenings at either end, depending on the surrounding conditions.

Cell Wall

The cell wall is distinctive, with a predominance of mesodiaminopimelic acid in the murein wall and many repetitions of arabinogalactan, as well as corynemycolic acid (a mycolic acid with 22 to 26 carbon atoms), bound by disaccharide bonds called L-Rhap-(1 → 4)--D-GlcNAc-phosphate. These form a complex commonly seen in *Corynebacterium* species: the mycolyl-AG–peptidoglican (mAGP).

Culture

Corynebacteria grow slowly, even on enriched media. In terms of nutritional requirements, all need biotin to grow. Some strains also need thiamine and PABA. Some of the *Corynebacterium* species with sequenced genomes have between 2.5 and 3.0 million base pairs. The bacteria grow in Loeffler's medium, blood agar, and trypticase soy agar (TSA). They form small, grayish colonies with a granular appearance, mostly translucent, but with opaque centers, convex, with continuous borders. The color tends to be yellowish-white in Loeffler's medium. In TSA, they can form grey colonies with black centers and dentated borders that look similar to flowers (*C. gravis*), or continuous borders (*C. mitis*), or a mix between the two forms (*C. intermedium*).

Habitat

Corynebacterium species occur commonly in nature in the soil, water, plants, and food products. The nondiphtheiroid *Corynebacterium* species can even be found in the mucosa and normal skin flora of humans and animals. Some species are known for their pathogenic effects in humans and other animals. Perhaps the most notable one is *C. diphtheriae*, which acquires the capacity to produce diphtheria toxin only after interacting with a bacteriophage. Other pathogenic species in humans include: *C. amicolatum*, *C. striatum*, *C. jeikeium*, *C. urealyticum*, and *C. xerosis*; all of these are important as pathogens in immunosuppressed patients. Pathogenic species in other animals include *C. bovis* and *C. renale*.

Role in Disease

The most notable human infection is diphtheria, caused by *C. diphtheriae*. It is an acute and contagious infection characterized by pseudomembranes of dead epithelial cells, white blood cells, red blood cells, and fibrin that form around the tonsils and back of the throat. In developed countries, it is an uncommon illness that tends to occur in unvaccinated individuals, especially school-aged children, elderly, neutropenic or immunocompromised patients, and those with prosthetic devices such as prosthetic heart valves, shunts, or catheters. It is more common in developing countries It can occasionally infect wounds, the vulva, the conjunctiva, and the middle ear. It can be spread within a hospital. The virulent and toxigenic strains are lysogenic, and produce an exotoxin formed by two polypeptide chains, which is itself produced when a bacterium is transformed by a gene from the β prophage.

Several species cause disease in animals, most notably *C. pseudotuberculosis*, which causes the disease caseous lymphadenitis, and some are also pathogenic in humans. Some attack healthy hosts, while others tend to attack the immunocompromised. Effects of infection include granulomatous lymphadenopathy, pneumonitis, pharyngitis, skin infections, and endocarditis. Corynebacterial endocarditis is seen most frequently in patients with intravascular devices. *C. tenuis* is believed to cause trichomycosis palmellina and trichomycosis axillaris. *C. striatum* may cause axillary odor. *C. minutissimum* causes erythrasma.

Industrial Uses

Nonpathogenic species of *Corynebacterium* are used for very important industrial applications, such as the production of amino acids, nucleotides, and other nutritional factors (Martín, 1989); bioconversion of steroids; degradation of hydrocarbons; cheese aging; and production of enzymes (Khurana et al., 2000). Some species produce metabolites similar to antibiotics: bacteriocins of the corynecin-linocin type, antitumor agents, etc. One of the most studied species is *C. glutamicum*, whose name refers to its capacity to produce glutamic acid in aerobic conditions. This is used in the food industry as monosodium glutamate in the production of soy sauce and yogurt.

Species of *Corynebacterium* have been used in the mass production of various amino acids including glutamic acid, a food additive that is made at a rate of 1.5 million tons/year. The metabolic pathways of *Corynebacterium* have been further manipulated to produce lysine and threonine.

L-Lysine production is specific to *C. glutamicum* in which core metabolic enzymes are manipulated through genetic engineering to drive metabolic flux towards the production of NADPH from the pentose phosphate pathway, and L-4-aspartyl phosphate, the commitment step to the synthesis of L-lysine, lysC, dapA, dapC, and dapF. These enzymes are up-regulated in industry through genetic engineering to ensure adequate amounts of lysine precursors are produced to increase metabolic flux. Unwanted side reactions such as threonine and asparagine production can occur if a buildup of intermediates occurs, so scientists have developed mutant strains of *C. glutamicum* through PCR engineering and chemical knockouts to ensure production of side-reaction enzymes are limited. Many genetic manipulations conducted in industry are by traditional cross-over methods or inhibition of transcriptional activators

Expression of functionally active human epidermal growth factor has been brought about in *C. glutamicum*, thus demonstrating a potential for industrial-scale production of human proteins. Expressed proteins can be targeted for secretion through either the general secretory pathway or the twin-arginine translocation pathway.

Unlike gram-negative bacteria, the gram-positive *Corynebacterium* species lack lipo-polysaccharides that function as antigenic endotoxins in humans.

Species

- *Corynebacterium efficiens*

Most species of corynebacteria are not lipophilic.

Nonlipophilic

The nonlipophilic bacteria may be classified as fermentative and nonfermentative:

- Fermentative corynebacteria

  - *Corynebacterium diphtheriae* group
  - *Corynebacterium xerosis* and *Corynebacterium striatum*
  - *Corynebacterium minutissimum*
  - *Corynebacterium amycolatum*
  - *Corynebacterium glucuronolyticum*
  - *Corynebacterium argentoratense*
  - *Corynebacterium matruchotii*
  - *Corynebacterium glutamicum*
  - *Corynebacterium* sp.

- Nonfermentative corynebacteria

  - *Corynebacterium afermentans* subsp. *afermentans*
  - *Corynebacterium auris*
  - *Corynebacterium pseudodiphtheriticum*
  - *Corynebacterium propinquum*

Lipophilic

- *Corynebacterium jeikeium*

- *Corynebacterium urealyticum*

- *Corynebacterium afermentans* subsp. *lipophilum*

- *Corynebacterium accolens*

- *Corynebacterium macginleyi*

- *CDC coryneform* groups F-1 and G

- *Corynebacterium bovis*

Novel Corynebacteria that do Not Contain Mycolic Acids

- *Corynebacterium kroppenstedtii*

Clostridium Botulinum

Clostridium botulinum is a Gram-positive, rod-shaped, anaerobic, spore-forming, motile bacterium with the ability to produce the neurotoxin botulinum. The botulinum toxin can cause a severe flaccid paralytic disease in humans and other animals and is the most potent toxin known to mankind, natural or synthetic, with a lethal dose of 1.3–2.1 ng/kg in humans.

C. botulinum is a diverse group of pathogenic bacteria initially grouped together by their ability to produce botulinum toxin and now known as four distinct groups, *C. botulinum* groups I-IV. *C. botulinum* groups I-IV, as well as some strains of *Clostridium butyricum* and *Clostridium baratii*, are the bacteria responsible for producing botulinum toxin.

C. botulinum is responsible for foodborne botulism (ingestion of preformed toxin), infant botulism (intestinal infection with toxin-forming *C. botulinum*), and wound botulism (infection of a wound with *C. botulinum*). *C. botulinum* produces heat-resistant endospores that are commonly found in soil and are able to survive under adverse conditions.

Microbiology

C. botulinum is a Gram-positive, rod-shaped, spore-forming bacterium. It is an obligate anaerobe, meaning that oxygen is poisonous to the cells. However, *C. botulinum* tolerates traces of oxygen due to the enzyme superoxide dismutase, which is an important antioxidant defense in nearly all cells exposed to oxygen. *C. botulinum* is only able to produce the neurotoxin during sporulation, which can only happen in an anaerobic environment. Other bacterial species produce spores in an unfavorable growth environment to preserve the organism's viability and permit survival in a dormant state until the spores are exposed to favorable conditions.

C. botulinum is divided into four distinct phenotypic groups (I-IV) and is also classified into seven serotypes (A-G) based on the antigenicity of the botulinum toxin produced.

Groups

The classification into groups is based on the ability of the organism to digest complex proteins. Studies at the DNA and rRNA level support the subdivision of the species into groups I-IV. Most outbreaks of human botulism are caused by group I (proteolytic) or II (non-proteolytic) *C. botulinum*. Group III organisms mainly cause diseases in animals. Group IV *C. botulinum* has not been shown to cause human or animal disease.

Botulinum Toxin

Neurotoxin production is the unifying feature of the species. Seven types of toxins have been identified that are allocated a letter (A-G). All toxins are rapidly destroyed at 100 °C, but they are resistant to degradation by enzymes found in the gastrointestinal tract. This allows for ingested toxin to be absorbed from the intestines into the bloodstream.

Most strains produce one type of neurotoxin, but strains producing multiple toxins have been described. *C. botulinum* producing B and F toxin types have been isolated from human botulism cases in New Mexico and California. The toxin type has been designated Bf as the type B toxin was found in excess to the type F. Similarly, strains producing Ab and Af toxins have been reported. Evidence indicates the neurotoxin genes have been the subject of horizontal gene transfer, possibly from a viral source. This theory is supported by the presence of integration sites flanking the toxin in some strains of *C. botulinum*. However, these integrations sites are degraded, indicating that the *C. botulinum* acquired the toxin genes quite far in the evolutionary past.

Botulinum Toxin Types

Only botulinum toxin types A, B, E, and F cause disease in humans. Types A, B, and E are associated with foodborne illness, with type E specifically associated with fish products. Type C produces limberneck in birds and type D causes botulism in other mammals. No disease is associated with type G. The "gold standard" for determining toxin type is a mouse bioassay, but the genes for types A, B, E, and F can now be readily differentiated using quantitative PCR.

A few strains from organisms genetically identified as other *Clostridium* species have caused human botulism: *C. butyricum* has produced type E toxin and *C. baratii* had produced type F toxin. The ability of *C. botulinum* to naturally transfer neurotoxin genes to other clostridia is concerning, especially in the food industry, where preservation systems are designed to destroy or inhibit only *C. botulinum* but not other *Clostridium* species.

Laboratory Isolation

In the laboratory, *C. botulinum* is usually isolated in tryptose sulfite cycloserine (TSC) growth medium in an anaerobic environment with less than 2% oxygen. This can be achieved by several commercial kits that use a chemical reaction to replace O_2 with CO_2. *C. botulinum* is a lipase-positive microorganism that grows between pH of 4.8 and 7.0 and cannot use lactose as a primary carbon source, characteristics important for biochemical identification.

Taxonomy History

C. botulinum was first recognized and isolated in 1895 by Emile van Ermengem from home-cured ham implicated in a botulism outbreak. The isolate was originally named *Bacillus botulinus*, after the Latin word for sausage, *botulus*. ("Sausage poisoning" was a common problem in 18th- and 19th-century Germany, and was most likely caused by botulism) However, isolates from subsequent outbreaks were always found to be anaerobic spore formers, so Ida A. Bengtson proposed that the organism be placed into the genus *Clostridium*, as the *Bacillus* genus was restricted to aerobic spore-forming rods.

Since 1959, all species producing the botulinum neurotoxins (types A-G) have been designated *C. botulinum*. Substantial phenotypic and genotypic evidence exists to demonstrate heterogeneity within the species. This has led to the reclassification of *C. botulinum* type G strains as a new species, *C. argentinense*.

Group I *C. botulinum* strains that do not produce a botulin toxin are referred to as *C. sporogenes*.

The complete genome of *C. botulinum* has been sequenced at Wellcome Trust Sanger Institute in 2007.

Pathology

Botulism poisoning can occur due to preserved or home-canned, low-acid food that was not processed using correct preservation times and/or pressure.

Foodborne botulism "Signs and symptoms of foodborne botulism typically begin between 18 and 36 hours after the toxin gets into your body, but can range from a few hours to several days, depending on the amount of toxin ingested."

- Double vision
- Blurred vision
- Drooping eyelid
- Nausea, vomiting, and abdominal cramps
- Slurred speech
- Trouble breathing
- Difficulty in swallowing

- Dry mouth
- Muscle weakness
- Constipation
- Reduced or absent deep tendon reactions, such as in the knee.

Wound botulism Most people who develop wound botulism inject drugs several times a day, so it's difficult to determine how long it takes for signs and symptoms to develop after the toxin enters the body. Most common in people who inject black tar heroin, wound botulism signs and symptoms include:

- Difficulty swallowing or speaking
- Facial weakness on both sides of the face
- Blurred or double vision
- Drooping eyelids
- Trouble breathing
- Paralysis

Infant botulism If infant botulism is related to food, such as honey, problems generally begin within 18 to 36 hours after the toxin enters the baby's body. Signs and symptoms include:

- Constipation (often the first sign)
- Floppy movements due to muscle weakness and trouble controlling the head
- Weak cry
- Irritability
- Drooling
- Drooping eyelids
- Tiredness
- Difficulty sucking or feeding
- Paralysis

Beneficial effects of botulinum toxin: Purified botulinum toxin is diluted by a physician for treatment:

- Congenital pelvic tilt
- Spasmodic dysphasia (the inability of the muscles of the larynx)

- Achalasia (esophageal stricture)

- Strabismus (crossed eyes)

- Paralysis of the facial muscles

- Failure of the cervix

- Blinking frequently

- Anti-cancer drug delivery

C. botulinum in Different Geographical Locations

A number of quantitative surveys for *C. botulinum* spores in the environment have suggested a prevalence of specific toxin types in given geographic areas, which remain unexplained.

North America

Type A *C. botulinum* predominates the soil samples from the western regions, while type B is the major type found in eastern areas. The type-B organisms were of the proteolytic type I. Sediments from the Great Lakes region were surveyed after outbreaks of botulism among commercially reared fish, and only type E spores were detected. In a survey, type-A strains were isolated from soils that were neutral to alkaline (average pH 7.5), while type-B strains were isolated from slightly acidic soils (average pH 6.25).

Europe

C. botulinum type E is prevalent in aquatic sediments in Norway and Sweden, Denmark, the Netherlands, the Baltic coast of Poland, and Russia. The type-E *C. botulinum* was suggested to be a true aquatic organism, which was indicated by the correlation between the level of type-E contamination and flooding of the land with seawater. As the land dried, the level of type E decreased and type B became dominant.

In soil and sediment from the United Kingdom, *C. botulinum* type B predominates. In general, the incidence is usually lower in soil than in sediment. In Italy, a survey conducted in the vicinity of Rome found a low level of contamination; all strains were proteolytic *C. botulinum* types A or B.

Australia

C. botulinum type A was found to be present in soil samples from mountain areas of Victoria. Type-B organisms were detected in marine mud from Tasmania. [Needs Source Checked] Type-A *C. botulinum* has been found in Sydney suburbs and types A and B were isolated from urban areas. In a well-defined area of the Darling-Downs region of Queensland, a study showed the prevalence and persistence of *C. botulinum* type B after many cases of botulism in horses.

Other

A "mouse protection" or "mouse bioassay" test determines the type of *C. botulinum* toxin present using monoclonal antibodies. An enzyme-linked immunosorbent assay (ELISA) with digoxigenin-labeled antibodies can also be used to detect the toxin, and quantitative PCR can detect the toxin genes in the organism.

C. botulinum is also used to prepare the medicaments Botox, Dysport, Xeomin, and Neurobloc used to selectively paralyze muscles to temporarily relieve muscle function. It has other "off-label" medical purposes, such as treating severe facial pain, such as that caused by trigeminal neuralgia.

Botulin toxin produced by *C. botulinum* is often believed to be a potential bioweapon as it is so potent that it takes about 75 nanograms to kill a person (LD_{50} of 1 ng/kg, assuming an average person weighs ~75 kg); 1 kilogram of it would be enough to kill the entire human population. For comparative purposes, a quarter of a typical grain of sand's weight (350 ng) of botulinum toxin would constitute a lethal dose for humans.

C. botulinum is a soil bacterium. The spores can survive in most environments and are very hard to kill. They can survive the temperature of boiling water at sea level, thus many foods are canned with a pressurized boil that achieves even higher temperatures, sufficient to kill the spores.

Growth of the bacterium can be prevented by high acidity, high ratio of dissolved sugar, high levels of oxygen, very low levels of moisture, or storage at temperatures below 3 °C (38 °F) for type A. For example, in a low-acid, canned vegetable such as green beans that are not heated enough to kill the spores (i.e., a pressurized environment) may provide an oxygen-free medium for the spores to grow and produce the toxin. However, pickles are sufficiently acidic to prevent growth; even if the spores are present, they pose no danger to the consumer. Honey, corn syrup, and other sweeteners may contain spores, but the spores cannot grow in a highly concentrated sugar solution; however, when a sweetener is diluted in the low-oxygen, low-acid digestive system of an infant, the spores can grow and produce toxin. As soon as infants begin eating solid food, the digestive juices become too acidic for the bacterium to grow.

Prevention Methods

The control of food-borne botulism caused by *C. botulinum* is based almost entirely on thermal destruction (heating) of the spores or inhibiting spore germination into bacteria and allowing cells to grow and produce toxins in foods. To prevent foodborne botulism:

- Use approved heat processes for commercially and home-canned foods (i.e., pressure-can low-acid foods such as corn or green beans, meat, or poultry).

- Discard all swollen, gassy, or spoiled canned foods. Double bag the cans or jars with plastic bags that are tightly closed. Then place the bags in a trash receptacle for non-recyclable trash outside the home. Keep it out of the reach of humans and pets.

- Do not taste or eat foods from containers that are leaking, have bulges or are swollen, look damaged or cracked, or seem abnormal in appearance. Do not use products that spurt liquid or foam when the container is opened.

- Boil home-processed, low-acid canned foods for 10 minutes prior to serving. For higher altitudes, add 1 minute for each 1,000 feet of elevation.

- Refrigerate all leftovers and cooked foods within 2 hours after cooking (1 hour if the temperature is above 90 °F).

- One of the most common causes of foodborne botulism is improperly home-canned food, especially low-acid foods such as vegetables and meats. Only a pressure cooker/canner allows water to reach 240 to 250 °F, a temperature that can kill the spores.

- Follow strict hygienic procedures to reduce contamination for at home canning. Use of pressure canners and cookers as recommended by the US Department of Agriculture. Keep oils infused with garlic or herbs refrigerated. Keep potatoes which have been baked while wrapped in aluminum foil hot until served or refrigerated. Boil home-processed, low-acid and tomato foods canned foods in a saucepan for 10 minutes before serving, even if you detect no signs of spoilage.

Growth Conditions

Conditions conducive of growth are dependent on various environmental factors. Growth of C. botulinum is a risk in low acid foods as defined by having a pH greater than 4.6 although growth is significantly retarded for pH below 4.9. There have been some cases and specific conditions reported to sustain growth with pH below 4.6.

Diagnostic Methods

In the U.S Physicians may consider the diagnosis if the patient's history and physical examination suggest botulism. However, these clues are usually not enough to allow a diagnosis of botulism. Other diseases such as Guillain-Barré syndrome, stroke, and myasthenia gravis can appear similar to botulism, and special tests may be needed to exclude these other conditions. These tests may include a brain scan, spinal fluid examination, nerve conduction test (electromyography, or EMG), and a tensilon test for myasthenia gravis. Tests for botulinum toxin and for bacteria that cause botulism can be performed at some state health department laboratories and at CDC.

Diplorickettsia Massiliensis

Diplorickettsia massiliensis species is an obligate intracellular, gram negative bacterium isolated from Ixodes ricinus ticks collected in Slovak republic forest geographically from southeastern part of Rovinka in 2006. They belong to the gammaproteobacteria class and are non endospore forming, small rods usually grouped in pairs. The bacteria are non-motile, and 16S rRNA, rpoB, parC and ftsY gene sequencing indicate that this bacterium is clearly different from all other recognized species. An initial phylogenetic analysis based on 16S rRNA, clustered *D. massiliensis* with *Rickettsiella grylli*. Because of its low 16S rDNA similarity (94%) with *R. grylli*, it was classified as a new genus *Diplorickettsia* into the family *Coxiellaceae* and the order *Legionellales*. *D. massiliensis* strain 20B was identified in three patients with suspected tick-borne infections that exhibited a specific seroconversion. The evidence of infection was further reconfirmed by using PCR-assay, thus established its role as a human pathogen and later whole genome sequencing was performed.

Description

Observation of *Diplorickettsia massiliensis* Strain 20B using Transmission electron microscopy

Diplorickettsia massiliensis (mas.si' li.en.sis. L. gen. adj. massiliensis, from Massilia, the Latin name of Marseille, France, where the organism was first grown, identified and characterized). The description is for that of genus type strain 20B. The known geographical distribution of this bacterium is Slovakia. This isolate has been deposited in the collection of the two World Health Organization Collaborative Centers for Rickettsial Reference and Research in Bratislava, Slovak Republic and the Faculté de Médecine, University of the Mediterranean in Marseille, France, as well as in the German Collection of Microorganisms and Cell Cultures (Deutsche Sammlung von Mikroorganismen und Zellkulturen, DSMZ) under the reference DSM 233381.

The GenBank/EMBL/DDBJ accession numbers for the 16S rRNA gene, rpoB, parC and ftsY sequences of strain B20 are GQ857549, GQ983049, GQ983050, and GU289825, respectively.

Phylogenetic Analysis

Comparative 16S rRNA gene sequence analysis showed that this strain belongs to the family *Coxiellaceae*, order *Legionellales* of *Gamma-proteobacteria*, and the closest relatives are different *Rickettsiella* spp. A 1476-bp fragment of the rrs gene encoding the 16S ribosomal RNA was amplified and sequenced. Surprisingly, a BLAST search did not result in a close similarity with any published 16S RNA gene sequences from other bacteria. The level of 16S rRNA gene sequence similarity between strain 20B and other recognized species of the family was below 94.5%. Partial sequences of the rpoB, parC and ftsY genes confirmed the phylogenetic position of the new isolate. The G+C content estimated on the basis of whole genome analysis of strain 20B was 37.88%. On the basis of its phenotypic and genotypic properties, together with phylogenetic distinctiveness, Mediannikov *et al.* proposed that strain 20B to be classified in the new genus *Diplorickettsia* as the type strain of a novel species named *Diplorickettsia massiliensis* sp. nov.

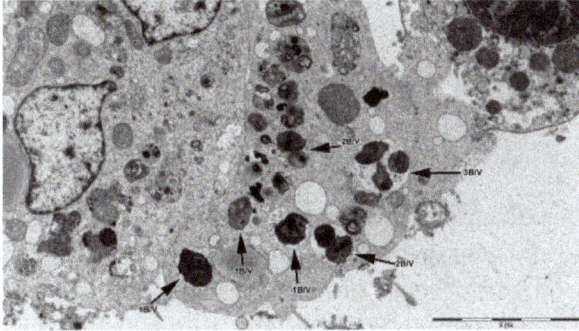

Diplorickettsia massiliensis Strain 20B bacteria grown in XTC-2 cells

Culture and Staining Observations

Bacteria visualized in a rich culture established in XTC-2cell line by Gimenez staining appeared as intracellular red rods usually grouped in pairs but not connected with each other. Manual counting of bacteria in an unlysed eukaryotic cell showed that almost all bacteria (97%) were paired. The bacteria accumulated in the cytoplasm of cells but not in the nucleus. The maximum number of bacteria observed in one cell numbered over 100. Infected cells were often disrupted during centrifugation using a Cytospin (Thermo Shandon) centrifuge as revealed by subsequent staining. The isolate was Gram-negative when extracellular bacteria were stained.

Diplorickettsia massiliensis Strain 20B visualization using Transmission electron microscopy with uranyl acetate staining

In addition, Mediannikov *et al.* have examined the percentage of infected cells and the mean number of bacteria per cell in different cell lines. The highest growth speed and the presence of a cytopathogenic effect when bacteria were cultivated in XTC-2 cells (*Xenopus laevis*). A cytopathogenic effect, including cellular layer detachment and cell disruption, was observed 3–5 days after inoculation. It was the only cell line with 100% cells infected the mean number of bacteria per cell was more than 100 in all studied series. The growth speed and bacteria accumulation in cells were lower in cell lines of human origin (HEL and MRC5) cultivated at 32 °C but were minimal in mouse L929 cells.

Multiple attempts at cultivation of the bacteria in solid axenic media (both common and Legionella-specific) were not successful.

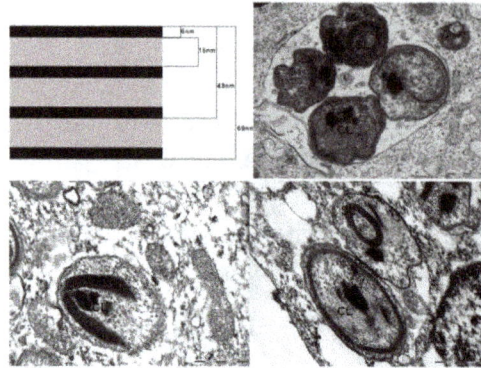

Internal structure of the bacteria using Transmission electron microscopy

Morphology by Electron Microscopy

Negative staining showed that the bacteria have an average length of 1540 nm (range: 848 to 3067 nm) and an average diameter of 695 nm (range: 515 to 992 nm). The longest bacteria were in the process of division. The bacteria of the strain 20B were harvested when they were extracellular for the studies of surface structures. Unlike other intracellular bacteria, including rickettsiae, 20B strain failed to highlight surface glycoproteins when colored with ruthenium red.

All bacteria observed intracellularly were located in vacuoles. Unlike *Rickettsiella*, they do not present a regular organization suggestive of crystalline structure. Based on visual impression of paired bacteria, they have counted the number of bacteria in vacuoles across the section: 51.4% of vacuoles contained 2 bacteria, 13.2% contained 3 or 4 bacteria, 1.7% contained more than 4, and 33.7% contained 1 bacterium. The ultrathin section may pass across only one bacterium in a vacuole that actually contains multiple bacteria, so this may mean that a number of these pseudo-single bacteria may be also paired. Taking these data into consideration, it was concluded that most bacteria are paired inside vacuoles. Bacteria in the process of division within the vacuoles were also found.

The internal structure of the bacteria was atypical. Electron-dense crystal-like structures were identified to be located in the center of almost all bacteria, usually surrounded by multilayer sheath-like structures. These layers alternate with electron-dense bands (6 nanometers) and light bands (15 nanometers). Up to seven electron-dense layers may be found in a single bacterium.

Endophyte

An endophyte is an endosymbiont, often a bacterium or fungus, that lives within a plant for at least part of its life cycle without causing apparent disease. Endophytes are ubiquitous and have been found in all species of plants studied to date; however, most of the endophyte/plant relationships are not well understood. Endophytes are also known to occur within lichens and algae. Many economically important grasses (e.g., *Festuca* spp. and *Lolium* spp.) carry fungal endophytes in genus *Epichloë*, some of which may enhance host growth, nutrient acquisition and may improve the plant's ability to tolerate abiotic stresses, such as drought, and enhance resistance to insects, plant pathogens and mammalian herbivores.

Transmission

Endophytes may be transmitted either vertically (directly from parent to offspring) or horizontally (among individuals). Vertically transmitted fungal endophytes are typically considered clonal and transmit via fungal hyphae penetrating the embryo within the host's seeds (e.g., seed transmitting forms of *Epichloë*). Conversely, reproduction through asexual or sexual spores leads to horizontal transmission, where endophytes may spread between plants in a population or community. Some endophytes that frequently transmit vertically may also produce spores on plants that can be transmitted horizontally (e.g., *Epichloë festucae*). Some of the *Epichloë* endophytes have been found to produce a cryptic but infective conidial state on the surfaces of leaf blades. However, the extent to which endophytes rely on these cryptic conidia for horizontal transmission is still unknown. Some endophytic fungi are actually latent pathogens or saprotrophs that only become active and reproduce under specific environmental conditions or when their host plants are stressed or begin to senesce.

Endophyte-Host Interactions

Endophytes may benefit host plants by preventing pathogenic or parasitic organisms from colonizing them. Extensive colonization of the plant tissue by endophytes creates a "barrier effect", where the local endophytes outcompete and prevent pathogenic organisms from taking hold. Endophytes may also produce chemicals which inhibit the growth of competitors, including pathogenic organisms. Endophytes are also known to increase expression of defense-related genes in plants, making plants more resistant to many potential pathogens. Some fungal and bacterial endophytes have proven to increase plant growth and improve overall plant hardiness. The presence of fungal endophytes can cause higher rates of water loss in leaves. However, certain microbial endophytes may also help plants to tolerate biotic stress such as root herbivory or abiotic stresses, including salt, drought or heat stresses. Endophytes have also been shown to enhance plant development and increase nutrient (phosphorus and nitrogen) uptake into plants. Endophyte-related host benefits are common phenomena, and have been the focus of much research, particularly among the grass endophytes. In spite of the many reports of beneficial effects of endophytes it has come to be understood that the relationship between endophytes and hosts may be considered a balanced antagonism with both positive and negative effects on hosts depending on the environmental conditions. Redman et al. advanced the hypothesis of 'habitat adapted symbiosis' where plants are proposed to associate with particular endophytes that increase tolerance or resistance to the predominant biotic or abiotic stresses of their habitats. Fungal and bacterial endophytes may comprise functional communities in plants that increase a plant's capacity to survive and thrive in its habitat.

Endophytes for Medicinal and Industrial Applications

The wide range of compounds produced by endophytes have been shown to combat pathogens and even cancers in animals including humans. One notable endophyte with medicinal benefits to humans was discovered by Gary Strobel: *Pestalotiopsis microspora*, an endophytic fungus of *Taxus wallachiana* (Himalayan Yew) was found to produce taxol. Also it was found that the endophytic fungus *Aspergillus flavus* from *Solanum nigrum* can produce solamargine. Endophytes are also being investigated for roles in biofuels production. Inoculating plants with certain endophytes may provide increased disease or parasite resistance while others may possess metabolic processes that convert cellulose and other carbon sources into "myco-diesel" hydrocarbons and hydrocarbon derivatives.

Endophytes for Agricultural Applications

Among the many promising applications of endophytic microbes are those intended to increase agricultural use of endophytes to produce crops that grow faster and are more resistant and hardier than crops lacking endophytes. *Epichloë* endophytes are being widely used commercially in turf grasses to enhance the performance of the turf and its resistance to biotic and abiotic stresses. *Piriformospora indica* is an interesting endophytic fungus of the order Sebacinales, the fungus is capable of colonising roots and forming symbiotic relationship with many plants. *P. indica* symbiosis has been shown to increase crop yield for a variety of crops (barley, tomato, maize etc.) and provide a measure of protection against pathogens and abiotic stresses. Recent evidence suggests that communities of bacterial and fungal endophytes may work in functional consortia to promote growth and protect plants in natural populations; while plants in intensive cultivation may lose these defensive and growth promotional microbiome components. Some scientists propose that restoration of defensive and growth promotional endophytes in agricultural crops could result in reduction of agrochemical inputs to control pests and diseases and result in crops that would better tolerate droughts and other stresses. There is some evidence that some bacterial endophytes may establish symbiosis with both plants and animals. This raises the possibility that crops could one day be produced that carry probiotic endophytes to enhance human health.

The Search for Endophytes

It is speculated that there may be many thousands of endophytes useful to mankind but since there are few scientists working in this field, and since environmental contamination, deforestation and biodiversity loss are widespread, many endophytes might be permanently lost before their utility is explored.

Endophytic species are very diverse; only a small minority of existing endophytes have been characterized. A single plant organ (leaf, stem or root) of a plant can harbor many different species of endophytes, both bacterial and fungal. Additionally, some endophytic bacteria may live within endophytic fungi.

Endophytes can be identified in several ways, usually through amplifying and sequencing a small piece of DNA. Some endophytes can be cultured from a piece of their host plant in an appropriate growth medium. An important step in culturing endophytes is to surface disinfect plant tissues prior to placement on culture media. This kills epiphytic microbes, ensuring only growth of endophytic microbes. Not all endophytes can be cultured in this way, as shown by discovery of cryptic, unculturable endophyte species through DNA based analysis of leaf tissue. Some grass endophytes in genus *Epichloë* can be seen as intercellular sinuous strands of hyphae under the microscope following leaf sheath or culm tissue staining with aniline blue. Many endophytes do not sporulate when cultured. Since fungal identification by morphology is based primarily on spore-bearing structures, this fact makes visual identification of some endophytic cultures challenging.

Diversity of Fungal Endophytes

Fungal endophytes are generally from the phylum Ascomycota, though other phyla are represented. Some specific examples of which are found in orders Hypocreales and Xylariales of the Sordariomycetes (Pyrenomycetes) class. Additionally the class of Loculoascomycetes includes endophytes. Although endophytes may be diverse taxonomically, Rodriguez et al. classified fungal endophytes broadly into ecological categories or functional classes.

Diversity of Algal Endophytes

A number of endophytes are now known that grow within seaweeds and algae. One such example is Ulvella leptochaete, which has recently been discovered from host algae including Cladophora and Laurentia from India.

Diversity of Bacterial Endophytes

Bacterial endophytes may belong to a broad range of taxa, including α-Proteobacteria, β-Proteobacteria, γ-Proteobacteria, Firmicutes, Actinobacteria, etc... Bacterial endophytes have been found to become intracellular in root and shoot cells of many plants, with entry into cells in the meristems. In this intracellular form bacteria lose cell walls but continue to divide and metabolize. These wall-less intracellular forms of bacteria are called L-forms. Paungfoo-Lonhienne et al. observed the degradation of intracellular microbes within root cells and hypothesized that intracellular microbes may be a source of organic nutrients or vitamins for plants; they termed this process 'rhizophagy'.

Hfr Cell

A high-frequency recombination cell (Hfr cell) (also called an Hfr strain) is a bacterium with a conjugative plasmid (for example, the F-factor) integrated into its chromosomal DNA. The integration of the plasmid into the cell's chromosome is through homologous recombination. A conjugative plasmid capable of chromosome integration is also called an episome (a segment of DNA that can exist as a plasmid or become integrated into the chromosome). When conjugation occurs, Hfr cells are very efficient in delivering chromosomal genes of the cell into recipient F⁻ cells, which lack the episome.

History

The Hfr strain was first characterized by Luca Cavalli-Sforza. William Hayes also isolated another Hfr strain independently.

Transfer of Bacterial Chromosome by Hfr Cells

An Hfr cell can transfer a portion of the bacterial genome. Despite being integrated into the chromosomal DNA of the bacteria, the F factor of Hfr cells can still initiate conjugative transfer, without being excised from the bacterial chromosome first. Due to the F factor's inherent tendency to transfer itself during conjugation, the rest of the bacterial genome is dragged along with it. Therefore, unlike a normal F⁺ cell, Hfr strains will attempt to transfer their *entire* DNA through the mating bridge, in a fashion similar to the normal conjugation. Interestingly, in a typical conjugation, the recipient cell also becomes F⁺ after conjugation as it receives an entire copy of the F factor plasmid; but this is not the case in conjugation mediated by Hfr cells. Due to the large size of bacterial chromosome, it is very rare for the entire chromosome to be transferred into the F⁻ cell as time required is simply too long for the cells to maintain their physical contact. Therefore, as the conjugative transfer is not complete (the circular nature of plasmid and bacterial chromosome requires complete transfer for the F factor to be transferred as it may be cut in the middle), the recipient ⁻ cells do not receive the complete F factor sequence, and do not become F ⁺ due to its inability to form a sex pilus.

Interrupted Mating Technique

In conjugation mediated by Hfr cells, transfer of DNA starts at the origin of transfer (*oriT*) located within the F factor and then continues clockwise or counterclockwise depending on the orientation of F factor in the chromosome. Therefore, the length of chromosomal DNA transferred into the F⁻ cell is proportional to the time that conjugation is allowed to happen. This results in sequential transfer of genes on the bacterial chromosome. Bacterial geneticists make use of this principle to map the genes on the bacterial chromosome. This technique is called interrupted mating as geneticists allow conjugation to take place for different periods of time before stopping conjugation with a high-speed blender. By using Hfr and F⁻ strains with one strain carrying mutations in several genes, each affecting a metabolic function or causing antibiotic resistance, and examining the phenotype of the recipient cells on selective agar plates, one can deduce which genes are transferred into the recipient cells first and therefore are closer to the *oriT* sequence on the chromosome.

Indicator Bacteria

Indicator bacteria are types of bacteria used to detect and estimate the level of fecal contamination of water. They are not dangerous to human health but are used to indicate the presence of a health risk.

Each gram of human feces contains approximately ~100 billion (1×10^{11}) bacteria. These bacteria may include species of pathogenic bacteria, such as *Salmonella* or *Campylobacter*, associated with gastroenteritis. In addition, feces may contain pathogenic viruses, protozoa and parasites. Fecal material can enter the environment from many sources including waste water treatment plants, livestock or poultry manure, sanitary landfills, septic systems, sewage sludge, pets and wildlife. If sufficient quantities are ingested, fecal pathogens can cause disease. The variety and often low concentrations of pathogens in environmental waters makes them difficult to test for individually. Public agencies therefore use the presence of other more abundant and more easily detected fecal bacteria as indicators of the presence of fecal contamination.

Criteria for Indicator Organisms

The US Environmental Protection Agency (EPA) lists the following criteria for an organism to be an ideal indicator of fecal contamination:

1. The organism should be present whenever enteric pathogens are present

2. The organism should be useful for all types of water

3. The organism should have a longer survival time than the hardiest enteric pathogen

4. The organism should not grow in water

5. The organism should be found in warm-blooded animals' intestines.

None of the types of indicator organisms that are currently in use fit all of these criteria perfectly, however, when cost is considered, use of indicators becomes necessary.

Types of Indicator Organisms

Commonly used indicator bacteria include total coliforms, or a subset of this group, fecal coliforms, which are found in the intestinal tracts of warm blooded animals. Total coliforms were used as fecal indicators by public agencies in the US as early as the 1920s. These organisms can be identified based on the fact that they all metabolize the sugar lactose, producing both acid and gas as byproducts. Fecal coliforms are more useful as indicators in recreational waters than total coliforms which include species that are naturally found in plants and soil; however, there are even some species of fecal coliforms that do not have a fecal origin, such as *Klebsiella pneumoniae*. Perhaps the biggest drawback to using coliforms as indicators is that they can grow in water under certain conditions.

Escherichia coli (*E. coli*) and enterococci are also used as indicators.

Current Methods of Detection

Membrane Filtration and Culture on Selective Media

Indicator bacteria can be cultured on media which are specifically formulated to allow the growth of the species of interest and inhibit growth of other organisms. Typically, environmental water samples are filtered through membranes with small pore sizes and then the membrane is placed onto a selective agar. It is often necessary to vary the volume of water sample filtered in order to prevent too few or too many colonies from forming on a plate. Bacterial colonies can be counted after 24 to 48 hours depending on the type of bacteria. Counts are reported as colony forming units per 100 mL (cfu/100 mL).

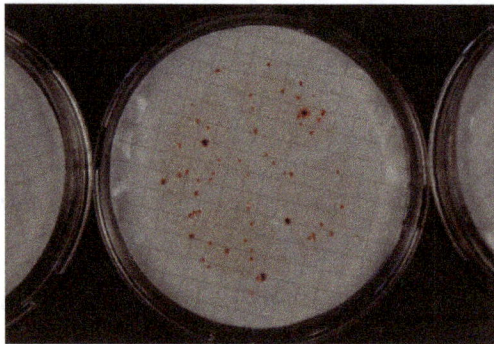

Enterococci colonies growing on a selective agar after membrane filtration.

Fast Detections Using Chromogenic Substances

One technique for detecting indicator organisms is the use of chromogenic compounds, which are added to conventional or newly devised media used for isolation of the indicator bacteria. These chromogenic compounds are modified to change color or fluorescence by the addition of either enzymes or specific bacterial metabolites. This enables for easy detection and avoids the need for isolation of pure cultures and confirmatory tests.

Application of Antibodies

Immunological methods using monoclonal antibodies can be used to detect indicator bacteria in water samples. Precultivation in select medium must preface detection to avoid detection of dead cells. ELISA antibody technology has been developed to allow for readable detection by the naked eye for rapid identification of coliform microcolonies. Other uses of antibodies in detection use magnetic beads coated with antibodies for the concentration and separation of the oocysts and cysts as described below for immunomagnetic separationz (IMS) methods.

IMS/Culture and Other Rapid Culture-based Methods

Immunomagnetic separation involves purified antigens biotinylated and bound to streptoavidin-coated paramagnetic particles. The raw sample is mixed with the beads, then a specific magnet is used to hold the target organisms against the vial wall and the non-bound material is poured off. This method can be used to recover specific indicator bacteria.

Gene Sequence-based Methods

Gene sequence-based methods depend on the recognition of exclusive gene sequences particular to specific strains of organisms. Polymerase chain reaction (PCR) and fluorescence in situ hybridization (FISH) are gene sequence-based methods currently being used to detect specific strains of indicator bacteria.

Water Quality Standards for Bacteria

Drinking Water Standards

World Health Organization Guidelines for Drinking Water Quality state that as an indicator organism *Escherichia coli* provides conclusive evidence of recent fecal pollution and should *not* be present in water meant for human consumption. In the U.S., the EPA *Total Coliform Rule* states that a water system is out of compliance if more than 5 percent of its monthly water samples contain coliforms.

Recreational Standards

Early studies showed that individuals who swam in waters with geometric mean coliform densities above 2300/100 mL for three days had higher illness rates. In the 1960s, these numbers were converted to fecal coliform concentrations assuming 18 percent of total coliforms were fecal. Consequently, the National Technical Advisory Committee in the US recommended the following standard for recreational waters in 1968: 10 percent of total samples during any 30-day period should not exceed 400 fecal coliforms/100 mL or a log mean of 200/100 mL (based on a minimum of 5 samples taken over not more than a 30-day period).

Despite criticism, EPA recommended this criterion again in 1976, however, the Agency initiated numerous studies in the 1970s and 1980s to overcome the weaknesses of the earlier studies. In 1986, EPA revised its bacteriological ambient water quality criteria recommendations to include *E. coli* and enterococci.

Water Type	Indicator	Acceptable Swimming-Associated Gastroenteritis Rate per 1000 Swimmers	Steady State Geometric Mean Indicator Density per 100 mL	Single Sample Maximum Allowable Density per 100 mL			
				Designated Beach Area (upper 75% C.L.)	Moderate Full Body Contact Recreation (upper 82% C.L.)	Lightly Used Full Body Contact Recreation (upper 90% C.L.)	Infrequently Used Full Body Contact Recreation (upper 95% C.L.)
Freshwater	*E. coli*	8	126	235	298	409	575
	enterococci	8	33	61	78	107	151
Marine Water	*E. coli*	19	35	104	158	276	501

Canada's National Agri-Environmental Standards Initiative's approach to characterizing risks associated with fecal water pollution bacterial water quality at agricultural sites is to compare these sites with those at reference sites away from human or livestock sources. This approach generally results in lower levels if *E. coli* being used as a standard or "benchmark" based on a study that indicated pathogens were detected in 80% of water samples with less than 100 cfu *E. coli* per 100 mL.

Risk Assessment for Exposure to Pathogens in Recreational Waters

The New River as it enters California is dark green, white (foam), and milky brown/green. Fecal coliforms and fecal streptococci have been consistently detected in the New River at the Mexico-US border.

Most cases of bacterial gastroenteritis are caused by food-borne enteric microorganisms, such as *Salmonella* and *Campylobacter*; however, it is also important to understand the risk of exposure to pathogens via recreational waters. This is especially the case in watersheds where human or animal wastes are discharged to streams and downstream waters are used for swimming or other recreational activities. Other important pathogens other than bacteria include viruses such as rotavirus, hepatitis A and hepatitis E and protozoa like giardia, cryptosporidium and *Naegleria fowleri*. Due to the difficulties associated with monitoring pathogens in the environment, risk assessments often rely on the use of indicator bacteria.

Epidemiological Studies

In the 1950s, a series of epidemiological studies were done in the US to determine the relationship between water quality of natural waters and the health of bathers. The results indicated that swimmers were more likely to have gastrointestinal symptoms, eye infections, skin complaints, ear, nose, and throat infections and respiratory illness than non-swimmers and in some cases, higher coliform levels correlated to higher incidence of gastrointestinal illness, although the sample sizes in these studies were small. Since then, studies have been done to confirm causative relations between swimming and certain health outcomes. A review of 22 studies in 1998 confirmed that the health risks for swimmers increased as the number of indicator bacteria increased in recreational waters and that *E. coli* and enterococci concentrations correlated best with health outcomes among all the indicators studied. The relative risk (RR) of illness for swimmers in polluted freshwater versus swimmers in unpolluted water was between 1-2 for the majority of the data sets reviewed. The same study concluded that bacterial indicators were not well correlated to virus concentrations.

Fate and Transport of Pathogens

Survival of pathogens in waste materials, soil, or water, depends on many environmental factors including temperature, pH, organic matter content, moisture, exposure to light, and the presence of other organisms. Fecal material can be directly deposited, washed into waters by overland runoff, transported through the ground, or discharged to surface waters via sewer lines, pipes, or drainage tiles. Risk of exposure to humans requires: (1) pathogens to survive and be present; (2) pathogens to recreate in surface waters; and (3) individuals to come in contact with water for sufficient time, or ingest sufficient volumes of water to receive an infectious dose. Die-off rates of bacteria in the environment are often exponential, therefore, direct deposition of fecal material into waters generally contribute higher concentrations of pathogens than material that must be transported overland or through the subsurface.

Human Exposure

In general, children, the elderly, and immunocompromised individuals require a lower dose of a pathogenic organism in order to contract an infection. Presently there are very few studies which are able to quantify the amount of time people are likely to spend in recreational waters and how much water they are likely to ingest. In general, children swim more often, stay in the water longer, submerge their heads more often, and swallow more water. This makes people more fearful of water in the sea as more bacteria will be growing on and around them.

Quantitative Microbiological Risk Assessment

Quantitative microbiological risk assessments (QMRAs) combine pathogen concentrations in water with dose-response relationships and data reflecting potential exposure to estimate the risk of infection.

Data on water exposure are generally collected using questionnaires, but may also be determined from actual measurements of water ingested, or estimated from previously published data. Respondents are asked to report the frequency and timing and location of exposures, detailed information about the amount of water swallowed and head submersion, and basic demographic characteristics such as age, gender, socioeconomic status and family composition. Once sufficient data are collected and determined to be representative of the general population, they are usually fit with distributions, and these distribution parameters are then used in the risk assessment equations. Monitoring data representing occurrence of pathogens, direct measurement of pathogen concentrations, or estimations deriving pathogen concentrations from indicator bacteria concentrations, are also fit with distributions. Dose is calculated by multiplying the concentration of pathogens per volume by volume. Dose-responses can also be fit with a distribution.

Risk Management and Policy Implications

The more assumptions that are made, the more uncertain estimates of risk related to pathogens will be. However, even with considerable uncertainty, QMRAs are a good way to compare different risk scenarios. In a study comparing estimated health risks from exposures to recreational waters impacted by human and non-human sources of fecal contamination, QMRA determined that the risk of gastrointestinal illness from exposure to waters impacted by cattle were similar to those impacted by human waste, and these were higher than for waters impacted by gull, chicken, or pig faeces. Such studies could be useful to risk managers for determining how best to focus their limited resources, however, risk managers must be aware of the limitations of data used in these calculations. For example, this study used data describing concentrations of *Salmonella* in chicken feces published in 1969. Methods for quantifying bacteria, changes in animal housing practices and sanitation, and many other factors may have changed the prevalence of *Salmonella* since that time. Also, such an approach often ignores the complicated fate and transport processes that determine bacteria concentrations from the source to the point of exposure.

Addressing Bacterial Water Quality Problems

In the US, individual states are allowed to develop their own water quality standards based on EPA's recommendations under the Clean Water Act of 1977. Once water quality standards are approved, states are tasked with monitoring their surface waters to determine where impairments occur, and watershed plans called Total Maximum Daily Loads (TMDLs) are developed to direct water quality improvement efforts including changes to allowable bacteria loading by point sources and recommendations for changes to practices that reduce nonpoint-source contributions to bacteria loads. Also, many states have beach monitoring programs to warn swimmers when high levels of indicator bacteria are detected.

Cyanobacteria

Cyanobacteria, also known as Cyanophyta, is a phylum of bacteria that obtain their energy through photosynthesis. The name "cyanobacteria" comes from the color of the bacteria. Sometimes, they are called blue-green algae, and incorrectly so, because cyanobacteria are prokaryotes and the term "algae" is reserved for eukaryotes.

Like other prokaryotes, cyanobacteria have no membrane-sheathed organelles. Photosynthesis is performed in distinctive folds in the outer membrane of the cell (unlike green plants which use organelles adapted for this specific role, called chloroplasts). Biologists commonly agree that chloroplasts found in eukaryotes, have their ancestry in cyanobacteria, via a process called endosymbiosis.

By producing oxygen as a byproduct of photosynthesis, cyanobacteria are thought to have converted the early oxygen-poor, reducing atmosphere, into an oxidizing one, causing the "rusting of the Earth" and the Great Oxygenation Event, that dramatically changed the composition of life forms and led to the near-extinction of anaerobic organisms.

Description

Cyanobacteria are a group of photosynthetic, nitrogen fixing bacteria that live in a wide variety of habitats such as moist soils and in water. They may be free-living or form symbiotic relationships with plants or with lichen-forming fungi as in the lichen genus *Peltigera*. They range from unicellular to filamentous and include colonial species. Colonies may form filaments, sheets, or even hollow balls. Some filamentous species can differentiate into several different cell types: vegetative cells, the normal, photosynthetic cells that are formed under favorable growing conditions; akinetes, climate-resistant spores that may form when environmental conditions become harsh; and thick-walled heterocysts, which contain the enzyme nitrogenase, vital for nitrogen fixation.

Nitrogen Fixation

Cyanobacteria can fix atmospheric nitrogen in anaerobic conditions by means of specialized cells called heterocysts. Heterocysts may also form under the appropriate environmental conditions (anoxic) when fixed nitrogen is scarce. Heterocyst-forming species are specialized for nitrogen fixation and are able to fix nitrogen gas into ammonia (NH_3), nitrites (NO_2^-) or nitrates (NO_3^-), which can be absorbed by plants and converted to protein and nucleic acids (atmospheric nitrogen is not bioavailable to plants, except for those having endosymbiotic nitrogen-fixing bacteria, especially the Fabaceae family, among others). Most importantly they are unicellular.

Free-living cyanobacteria are present in the water column in rice paddies, and cyanobacteria can be found growing as epiphytes on the surfaces of the green alga, Chara, where they may fix nitrogen. Cyanobacteria such as (*Anabaena*, a symbiont of the aquatic fern *Azolla*), can provide rice plantations with biofertilizer.

Morphology

Many cyanobacteria form motile filaments of cells, called hormogonia, that travel away from the main biomass to bud and form new colonies elsewhere. The cells in a hormogonium are often thinner than in the vegetative state, and the cells on either end of the motile chain may be tapered. To break away from the parent colony, a hormogonium often must tear apart a weaker cell in a filament, called a necridium.

Colonies of *Nostoc pruniforme*

Each individual cell of a cyanobacterium typically has a thick, gelatinous cell wall. They lack flagella, but hormogonia of some species can move about by gliding along surfaces. Many of the multicellular filamentous forms of *Oscillatoria* are capable of a waving motion; the filament oscillates back and forth. In water columns, some cyanobacteria float by forming gas vesicles, as in archaea. These vesicles are not organelles as such. They are not bounded by lipid membranes, but by a protein sheath.

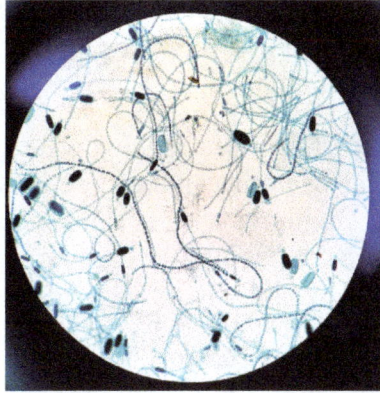

Cylindrospermum sp.

Ecology

Cyanobacteria can be found in almost every terrestrial and aquatic habitat—oceans, fresh water, damp soil, temporarily moistened rocks in deserts, bare rock and soil, and even Antarctic rocks. They can occur as planktonic cells or form phototrophic biofilms. They are found in almost every endolithic ecosystem. A few are endosymbionts in lichens, plants, various protists, or sponges and provide energy for the host. Some live in the fur of sloths, providing a form of camouflage.

Cyanobacterial bloom near Fiji

Aquatic cyanobacteria are known for their extensive and highly visible blooms that can form in both freshwater and marine environments. The blooms can have the appearance of blue-green paint or scum. These blooms can be toxic, and frequently lead to the closure of recreational waters when spotted. Marine bacteriophages are significant parasites of unicellular marine cyanobacteria.

Cyanobacteria prefer calm waters, such as those provided by ponds and lakes. Their life cycles are disrupted when the water naturally or artificially mixes from churning currents caused by the flowing water of streams or the churning water of fountains. For this reason blooms of cyanobacteria seldom occur in rivers unless the water is flowing slowly. When the bacteria are found in rivers, they have usually come from the outfall of lakes upstream from the sampling point.

Cyanobacteria are a growing concern for drinking water utilities who use lakes and rivers as their source water. The bacteria can interfere with treatment in various ways, primarily by plugging filters (often large beds of sand and similar media), and by producing cyanotoxins, which have the potential of causing serious illness if consumed.

Some of these organisms contribute significantly to global ecology and the oxygen cycle. The tiny marine cyanobacterium *Prochlorococcus* was discovered in 1986 and accounts for more than half of the photosynthesis of the open ocean. Many cyanobacteria even display the circadian rhythms that were once thought to exist only in eukaryotic cells.

"Cyanobacteria are arguably the most successful group of microorganisms on earth. They are the most genetically diverse; they occupy a broad range of habitats across all latitudes, widespread in freshwater, marine, and terrestrial ecosystems, and they are found in the most extreme niches such as hot springs, salt works, and hypersaline bays. Photoautotrophic, oxygen-producing cyanobacteria created the conditions in the planet's early atmosphere that directed the evolution of aerobic metabolism and eukaryotic photosynthesis. Cyanobacteria fulfill vital ecological functions in the world's oceans, being important contributors to global carbon and nitrogen budgets." – Stewart and Falconer

Photosynthesis

While contemporary cyanobacteria are linked to the plant kingdom as descendants of the endosymbiotic progenitor of the chloroplast, there are several features which are unique to this group.

Carbon Fixation

Cyanobacteria use the energy of sunlight to drive photosynthesis, a process where the energy of light is used to split water molecules into oxygen, protons, and electrons. Because they are aquatic organisms, they typically employ several strategies which are collectively known as a "carbon concentrating mechanism" to aid in the acquisition of inorganic carbon (CO_2 or bicarbonate). Among the more specific strategies is the widespread prevalence of the bacterial microcompartments known as carboxysomes. These icosahedral structures are composed of hexameric shell proteins that assemble into cage-like structures that can be several hundreds of nanometers in diameter. It is believed that these structures tether the CO_2-fixing enzyme, RuBisCO, to the interior of the shell, as well as the enzyme carbonic anhydrase, using the paradigm of metabolic channeling to enhance the local CO2 concentrations and thus increase the efficiency of the RuBisCO enzyme.

Electron Transport

In contrast to chloroplast-containing eukaryotes, cyanobacteria lack compartmental-ization of their thylakoid membranes, which are contiguous with the plasma membrane. Thus, the protein complexes involved in respiratory energy metabolism share several mobile energy carrier pools (e.g., the Quinone pool, cytochrome c, ferredoxins), so photosynthetic and respiratory metabolism interact with each other. Furthermore, there is a tremendous diversity among the respiratory components between species. Thus cyanobacteria can be said to have a "branched electron transport chain", analogous to the situation in purple bacteria

While most of the high-energy electrons derived from water are used by the cyanobacterial cells for their own needs, a fraction of these electrons may be donated to the external environment via electrogenic activity.

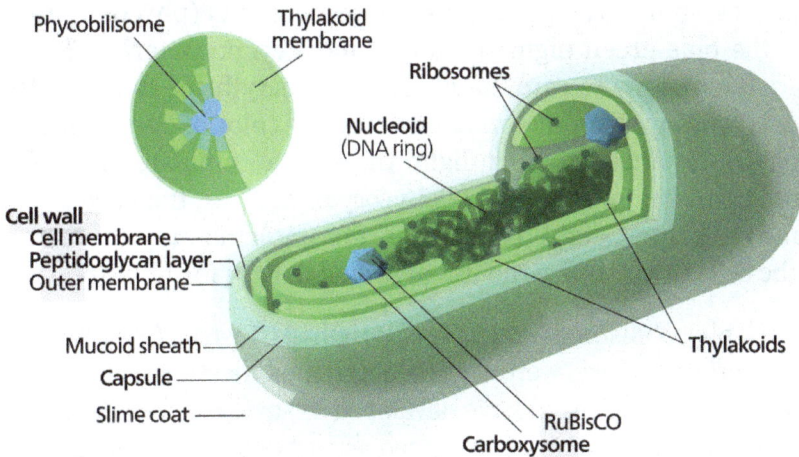

Metabolism and Organelles

As prokaryotes, cyanobacteria do not have nuclei or an internal membrane system. In most forms, the photosynthetic machinery is embedded into folds of the external cell membrane, called thylakoids. Cyanobacteria get their colour from the bluish pigment phycocyanin, which they use to capture light for photosynthesis. In general, photosynthesis in cyanobacteria uses water as an electron donor and produces oxygen as a byproduct, though some may also use hydrogen sulfide a process which occurs among other photosynthetic bacteria such as the purple sulfur bacteria. Carbon dioxide is reduced to form carbohydrates via the Calvin cycle.The large amounts of oxygen in the atmosphere are considered to have been first created by the activities of ancient cyanobacteria. They are often found as symbionts with a number of other groups of organisms such as fungi (lichens), corals, pteridophytes (*Azolla*), angiosperms (*Gunnera*), etc.

Many cyanobacteria are able to reduce nitrogen and carbon dioxide under aerobic conditions, a fact that may be responsible for their evolutionary and ecological success.

The water-oxidizing photosynthesis is accomplished by coupling the activity of photosystem (PS) II and I (Z-scheme). In anaerobic conditions, they are able to use only PS I—cyclic photophosphorylation—with electron donors other than water (hydrogen sulfide, thiosulphate, or even molecular hydrogen) just like purple photosynthetic bacteria. Furthermore, they share an archaeal property, the ability to reduce elemental sulfur by anaerobic respiration in the dark. Their photosynthetic electron transport shares the same compartment as the components of respiratory electron transport. Their plasma membrane contains only components of the respiratory chain, while the thylakoid membrane hosts an interlinked respiratory and photosynthetic electron transport chain. The terminal oxidases in the thylakoid membrane respiratory/photosynthetic electron transport chain are essential for survival to rapid light changes, although not for dark maintenance under conditions where cells are not light stressed.

Attached to the thylakoid membrane, phycobilisomes act as light-harvesting antennae for the photosystems. The phycobilisome components (phycobiliproteins) are responsible for the blue-green pigmentation of most cyanobacteria. The variations on this theme are due mainly to carotenoids and phycoerythrins that give the cells their red-brownish coloration. In some cyanobacteria, the color of light influences the composition of phycobilisomes. In green light, the cells accumulate more phycoerythrin, whereas in red light they produce more phycocyanin. Thus, the bacteria appear green in red light and red in green light. This process of complementary chromatic adaptation is a way for the cells to maximize the use of available light for photosynthesis.

A few genera lack phycobilisomes and have chlorophyll b instead (*Prochloron*, *Prochlorococcus*, *Prochlorothrix*). These were originally grouped together as the prochlorophytes or chloroxybacteria, but appear to have developed in several different lines of cyanobacteria. For this reason, they are now considered as part of the cyanobacterial group.

There are some groups capable of heterotrophic growth, while others are parasitic, causing diseases in invertebrates or eukaryotic algae (e.g., the black band disease).

Relationship to Chloroplasts

Chloroplasts found in eukaryotes (algae and plants) appear to have evolved from an endosymbiotic relation with cyanobacteria. This endosymbiotic theory is supported by various structural and genetic similarities. Primary chloroplasts are found among the "true plants" or green plants – species ranging from sea lettuce to evergreens and flowers that contain chlorophyll *b* – as well as among the red algae and glaucophytes, marine species that contain phycobilins. It now appears that these chloroplasts probably had a single origin, in an ancestor of the clade called Archaeplastida. However this does not necessitate origin from cyanobacteria themselves; microbiology is still undergoing profound classification changes and entire domains (such as Archaea) are poorly mapped and understood. Other algae likely took their chloroplasts from these forms by secondary endosymbiosis or ingestion.

Classification

Historically, bacteria were first classified as plants constituting the class Schizomycetes, which along with the Schizophyceae (blue-green algae/Cyanobacteria) formed the phylum Schizophyta, then in the phylum Monera in the kingdom Protista by Haeckel in 1866, comprising *Protogens, Protamaeba, Vampyrella, Protomonae,* and *Vibrio*, but not *Nostoc* and other cyanobacteria, which were classified with algae, later reclassified as the *Prokaryotes* by Chatton.

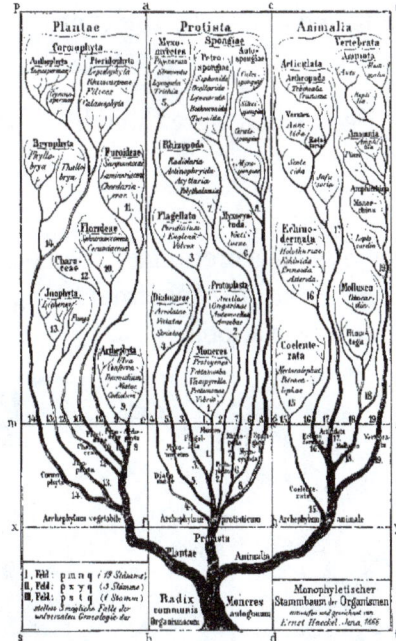

Tree of Life in *Generelle Morphologie der Organismen* (1866). Note the location of the genus *Nostoc* with algae and not with bacteria (kingdom "Monera")

The cyanobacteria were traditionally classified by morphology into five sections, referred to by the numerals I-V. The first three – Chroococcales, Pleurocapsales, and Oscillatoriales – are not supported by phylogenetic studies. The latter two – Nostocales and Stigonematales – are monophyletic, and make up the heterocystous cyanobacteria.

The members of Chroococcales are unicellular and usually aggregate in colonies. The classic taxonomic criterion has been the cell morphology and the plane of cell division. In Pleurocapsales, the cells have the ability to form internal spores (baeocytes). The rest of the sections include filamentous species. In Oscillatoriales, the cells are uniseriately arranged and do not form specialized cells (akinetes and heterocysts). In Nostocales and Stigonematales, the cells have the ability to develop heterocysts in certain conditions. Stigonematales, unlike Nostocales, include species with truly branched trichomes.

Most taxa included in the phylum or division Cyanobacteria have not yet been validly published under the Bacteriological Code, except:

- The classes Chroobacteria, Hormogoneae, and Gloeobacteria

- The orders Chroococcales, Gloeobacterales, Nostocales, Oscillatoriales, Pleuro-capsales, and Stigonematales

- The families Prochloraceae and Prochlorotrichaceae

- The genera *Halospirulina, Planktothricoides, Prochlorococcus, Prochloron,* and *Prochlorothrix*

The remainder are validly published under the International Code of Nomenclature for algae, fungi, and plants.

Formerly, some bacteria, like *Beggiatoa*, were thought to be colorless Cyanobacteria.

Earth History

Stromatolites are layered biochemical accretionary structures formed in shallow water by the trapping, binding, and cementation of sedimentary grains by biofilms (microbial mats) of microorganisms, especially cyanobacteria.

Stromatolites left behind by cyanobacteria are the oldest known fossils of life on Earth. This one-billion-year-old fossil is from Glacier National Park in Montana.

During the Precambrian, stromatolite communities of microorganisms grew in most marine and non-marine environments in the photic zone. After the Cambrian explosion of marine animals, grazing on the stromatolite mats by herbivores greatly reduced the occurrence of the stromatolites in marine environments. Since then, they are found mostly in hypersaline conditions where grazing invertebrates can't live (e.g. Shark Bay, Western Australia). Stromatolites provide ancient records of life on Earth by fossil remains which might date from more than 3.5 Ga ago, but this is disputed. As of 2010 the oldest undisputed evidence of cyanobacteria is from 2.1 Ga ago, but there is some evidence for them as far back as 2.7 Ga ago. Oxygen levels in the atmosphere remained around or below 1% of today's level until 2.4 Ga ago (the Great Oxygenation Event). The rise in oxygen may have caused a fall in methane levels, and triggered the Huronian glaciation from around 2.4 to 2.1 Ga ago. In this way, cyanobacteria may have killed off much of the other bacteria of the time.

Oncolites are sedimentary structures composed of oncoids, which are layered structures formed by cyanobacterial growth. Oncolites are similar to stromatolites, but instead of forming columns, they form approximately spherical structures that were not attached to the underlying substrate as they formed. The oncoids often form around a central nucleus, such as a shell fragment, and a calcium carbonate structure is deposited by encrusting microbes. Oncolites are indicators of warm waters in the photic zone, but are also known in contemporary freshwater environments. These structures rarely exceed 10 cm in diameter.

Biotechnology and Applications

The unicellular cyanobacterium *Synechocystis* sp. PCC6803 was the third prokaryote and first photosynthetic organism whose genome was completely sequenced. It continues to be an important model organism. *Cyanothece* ATCC 51142 is an important diazotrophic model organism. The smallest genomes have been found in *Prochlorococcus* spp. (1.7 Mb) and the largest in *Nostoc punctiforme* (9 Mb). Those of *Calothrix* spp. are estimated at 12–15 Mb, as large as yeast.

Oncolites from the Late Devonian Alamo bolide impact in Nevada

Recent research has suggested the potential application of cyanobacteria to the generation of renewable energy by converting sunlight into electricity. Internal photosynthetic pathways can be coupled to chemical mediators that transfer electrons to external electrodes. Currently, efforts are underway to commercialize algae-based fuels such as diesel, gasoline, and jet fuel.

Researchers from a company called Algenol have cultured genetically modified cyanobacteria in sea water inside a clear plastic enclosure so they first make sugar (pyruvate) from CO_2 and the water via photosynthesis. Then, the bacteria secrete ethanol from the cell into the salt water. As the day progresses, and the solar radiation intensifies, ethanol concentrations build up and the ethanol itself evaporates onto the roof of the enclosure. As the sun recedes, evaporated ethanol and water condense into droplets, which run along the plastic walls and into ethanol collectors, from where it is extracted from the enclosure with the water and ethanol separated outside the enclosure. As of March 2013, Algenol was claiming to have tested its technology in Florida and to have achieved yields of 9,000 US gallons per acre per year. This could potentially meet US demands for ethanol in gasoline in 2025, assuming a B30 blend, from an area of around half the size of California's San Bernardino County, requiring less than one-tenth of the area than ethanol from other biomass, such as corn, and only very limited amounts of fresh water.

Cyanobacteria cultured in specific media: Cyanobacteria can be helpful in agriculture as they have the ability to fix atmospheric nitrogen in soil.

Cyanobacteria may possess the ability to produce substances that could one day serve as anti-inflammatory agents and combat bacterial infections in humans.

Spirulina's extracted blue color is used as a natural food coloring in gum and candy.

Researchers from several space agencies argue that cyanobacteria could be used for producing goods for human consumption (food, oxygen...) in future manned outposts on Mars, by transforming materials available on this planet.

Health Risks

Cyanobacteria can produce neurotoxins, cytotoxins, endotoxins, and hepatotoxins (i.e. the microcystin-producing bacteria species microcystis), and are called cyanotoxins.

Specific toxins include, anatoxin-a, anatoxin-as, aplysiatoxin, cyanopeptolin, cylindrospermopsin, domoic acid, nodularin R (from *Nodularia*), neosaxitoxin, and saxitoxin. Cyanobacteria reproduce explosively under certain conditions. This results in algal blooms, which can become harmful to other species, and pose a danger to humans and animals, if the cyanobacteria involved produce toxins. Several cases of human poisoning have been documented, but a lack of knowledge prevents an accurate assessment of the risks.

Recent studies suggest that significant exposure to high levels of cyanobacteria producing toxins such as BMAA can cause amyotrophic lateral sclerosis (ALS). People living within a half-mile of cyanobacterially contaminated lakes have had a 2.3-times greater risk of developing ALS than the rest of the population; people around New Hampshire's Lake Mascoma had an up to 25 times greater risk of ALS than the expected incidence. BMAA from desert crusts found throughout Qatar might have contributed to higher rates of ALS in Gulf War veterans.

Chemical Control

Several chemicals can eliminate blue-green algal blooms from water-based systems. They include: calcium hypochlorite, copper sulphate, cupricide, and simazine. The calcium hypochlorite amount needed varies depending on the cyanobacteria bloom, and treatment is needed periodically. According to the Department of Agriculture Australia, a rate of 12 g of 70% material in 1000 l of water is often effective to treat a bloom. Copper sulfate is also used commonly, but no longer recommended by the Australian Department of Agriculture, as it kills livestock, crustaceans, and fish. Culpricide is a chelated copper product that eliminates blooms with lower toxicity risks than copper sulfate. Dosage recommendations vary from 190 ml to 4.8 l per 1000 m². Ferric alum treatments at the rate of 50 mg/l will reduce algae blooms. Simazine, which is also a herbicide, will continue to kill blooms for several days after an application. Simazine is marketed at different strengths (25, 50, and 90%), the recommended amount needed for one cubic meter of water per product is 25% product 8 ml; 50% product 4 ml; or 90% product 2.2 ml.

Dietary Supplementation

Some cyanobacteria are sold as food, notably *Aphanizomenon flos-aquae* and *Arthrospira platensis* (Spirulina).

Spirulina tablets

Despite the associated toxins which many of the members of this phylum produce, some microalgae also contain substances of high biological value, such as polyunsaturated fatty acids, amino acids (proteins), pigments, antioxidants, vitamins, and minerals. Edible blue-green algae reduce the production of pro-inflammatory cytokines by inhibiting NF-κB pathway in macrophages and splenocytes. Sulfate polysaccharides exhibit immunomodulatory, antitumor, antithrombotic, anticoagulant, anti-mutagenic, anti-inflammatory, antimicrobial, and even antiviral activity against HIV, herpes, and hepatitis.

References

- Tortora, Gerard (2010). Microbiology: An Introduction. San Francisco, CA: Benjamin Cummings. pp. 85–87, 161, 165,. ISBN 0-321-55007-2.

- Krieg, N. R.; Holt, J. G., eds. (1984). Bergey's Manual of Systematic Bacteriology. 1 (First ed.). Baltimore: The Williams & Wilkins Co. pp. 408–420. ISBN 0-683-04108-8.

- Brenner DJ, Krieg NR, Staley JT (July 26, 2005) [1984 (Williams & Wilkins)]. George M. Garrity, ed. The Gammaproteobacteria. Bergey's Manual of Systematic Bacteriology. 2B (2nd ed.). New York: Springer. p. 1108. ISBN 978-0-387-24144-9.

- Farrar J, Hotez P, Junghanss T, Kang G, Lalloo D, White NJ, eds. (2013). Manson's Tropical Diseases (23rd ed.). Oxford: Elsevier/Saunders. ISBN 9780702053061.

- Sieper, Joachim; Braun, Jürgen (2011). Ankylosing Spondylitis in Clinical Practice. London: Springer-Verlag. p. 9. ISBN 978-0-85729-179-0. Retrieved October 10, 2012.

- Cornelis P, ed. (2008). Pseudomonas: Genomics and Molecular Biology (1st ed.). Caister Academic Press. ISBN 1-904455-19-0.

- Krieg, Noel (1984). Bergey's Manual of Systematic Bacteriology, Volume 1. Baltimore: Williams & Wilkins. ISBN 0-683-04108-8.

- "Guidance for Commercial Processors of Acidified & Low-Acid Canned Foods". U.S. Food and Drug Administration. Retrieved 8 October 2016.

- "Biotechnology of endophytic fungi of grasses / edited by Charles W. Bacon, James F. White, Jr. - Version details - Trove". trove.nla.gov.au. Retrieved 2015-09-30.

- Garwood, Russell J. (2012). "Patterns In Palaeontology: The first 3 billion years of evolution". Palaeontology Online. 2 (11): 1–14. Retrieved June 25, 2015.

- "Facts about E. coli: dimensions, as discussed in bacteria: Diversity of structure of bacteria: – Britannica Online Encyclopedia". Britannica.com. Retrieved 2015-06-25.

- "Infections from some foodborne germs increased, while others remained unchanged in 2012" (Press release). CDC. April 18, 2013. Retrieved October 22, 2015.

- "Campylobacter survey: cumulative results from the full 12 months (Q1 – Q4)" (Press release). Food Standards Agency. May 28, 2015. Retrieved October 23, 2015.

Study of Protein in Bacteria

In order to study proteins in bacteria it is necessary to understand AB5 toxin, actin assembly-inducing protein, bacterial effector protein, cholera toxin and lac repressor. AB5 toxins have six component protein complexes and bacterial effectors are proteins secreted by pathogenic bacteria. This chapter serves as a source to understand bacteria in an explicated manner.

AB5 Toxin

The AB5 toxins are six-component protein complexes secreted by certain pathogenic bacteria known to cause human diseases such as cholera, dysentery, and hemolytic-uremic syndrome. One component is known as the A subunit, and the remaining five components make up the B subunit. All of these toxins share a similar structure and mechanism for entering targeted host cells. The B subunit is responsible for binding to receptors to open up a pathway for the A subunit to enter the cell. The A subunit is then able to use its catalytic machinery to take over the host cell's regular functions.

Families

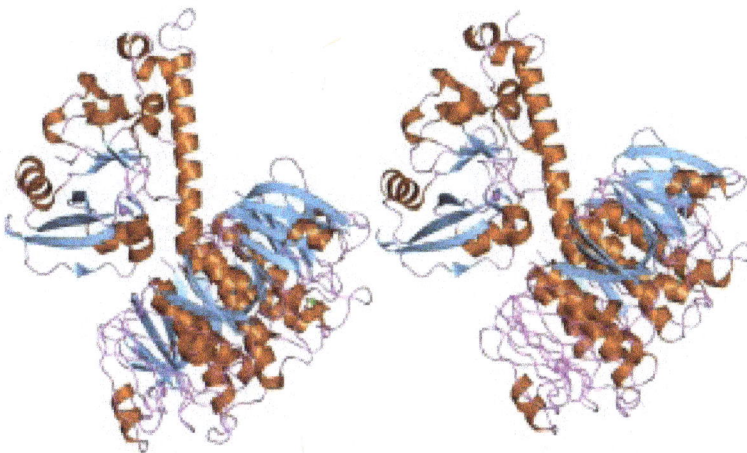

Ribbon diagram of cholera toxin. From PDB: 1s5e.

There are four main families of the AB5 toxin. These families are characterized by the sequence of their A subunit, as well as their catalytic ability.

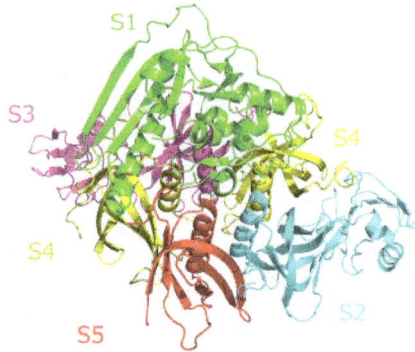

Ribbon diagram of pertussis toxin. S1 is the A subunit, and S2-S5 make up the B subunit.

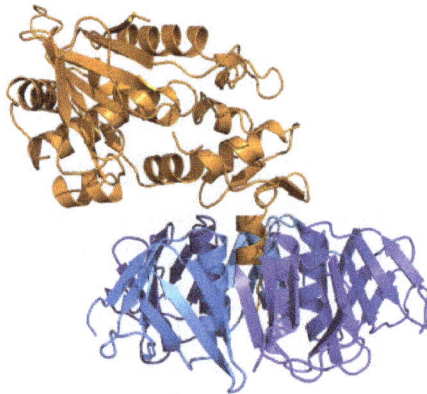

Ribbon diagram of shiga toxin (Stx) from *Shigella dysenteriae*, showing the characteristic AB5 structure. A subunit is in orange and B-subunit complex is in blue. From PDB: 1R4Q.

Cholera Toxin

This family is also known as Ct or Ctx, and includes the heat-labile enterotoxin family, known as LT. Cholera toxin's discovery is credited by many to Dr. Sambhu Nath De. He conducted his research in Calcutta (now Kolkata) making his discovery in 1959, although it was first purified by Robert Koch in 1883. Cholera toxin is an infectious toxin composed of a protein complex that is secreted by the bacterium *Vibrio cholerae*. Some symptoms of this toxin include chronic and widespread watery diarrhea and dehydration that, in some cases, leads to death.

Pertussis Toxin

This family is also known as Ptx and contains the toxin responsible for whooping cough. Pertussis toxin is secreted by the gram-negative bacterium, *Bordetella pertussis*. Whooping cough is very contagious and cases are slowly increasing in the United States despite vaccination. Symptoms include paroxysmal cough with whooping and even vomiting. The bacterium *Bordetella pertussis* was first identified as the cause of whooping cough and isolated by Jules Bordet and Octave Gengou in France in 1900.

Shiga Toxin

Shiga toxin is an infectious disease caused by the rod shaped *Shigella dysenteriae* as well as *Escherichia coli* (STEC), and is also known as Stx. Contaminated food and drinks are the source of infection and how this toxin is spread. Symptoms include abdominal pain as well as watery diarrhea. Severe life-threatening cases are characterized by hemorrhagic colitis (HC). The discovery of shiga toxin is credited to Dr. Kiyoshi Shiga in 1898.

Subtilase Cytotoxin

This family is also known as SubAB and was discovered during the 1990s. It produced by strains of STEC that do not have the locus of enterocyte effacement (LEE), and is known to cause hemolytic-uremic syndrome (HUS). It is called a subtilase cytotoxin because its A subunit sequence is similar to that of a subtilase-like serine protease in *Bacillus anthracis*. Some symptoms caused by this toxin are a decrease in platelet count in the blood or thrombocytopenia, an increase in white blood cell count or leukocytosis, and renal cell damage.

Structure

A complete AB5 toxin complex contains six protein units. Five units are similar or identical in structure and they comprise the B subunit. The last protein unit is unique and is known as the A subunit.

General diagram of the A subunit of the AB5 toxin with the disulfide linkage.

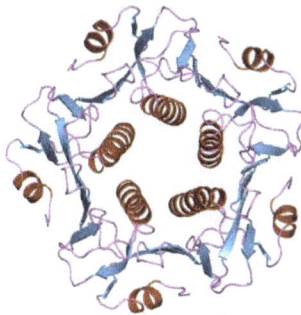

Ribbon diagram of the B-subunit of the cholera toxin.

A Subunit

The A subunit of an AB5 toxin is the portion responsible for catalysis of specific targets. For Shiga toxin family, the A subunit hosts a Trypsin-sensitive region which gives out two fragmented domains when cleaved. This region has not been confirmed for the other AB5 toxin families as yet. In general, the two domains of the A subunit, named A1 and A2, are linked by a disulfide bond. Domain A1 (approximately 22kDa in cholera toxin or heat labile enterotoxins) is the part of the toxin responsible for its toxic effects. Domain A2 (approximately 5kDa in cholera toxin or heat labile enterotoxin) provides a non-covalent linkage to the B subunit through the B subunit's central pore. The A1 chain for cholera toxin catalyzes the transfer of ADP-ribose from Nicotinamide adenine dinucleotide(NAD) to arginine or other guanidine compounds by utilizing ADP-ribosylation factors (ARFs). In the absence of arginine or simple guanidino compounds, the toxin mediated NAD+ nucleosidase (NADase) activity proceeds using water as a nucleophile.

B Subunit

The B subunits form a five-membered or pentameric ring, where one end of the A subunit goes into and is held. This B subunit ring is also capable of binding to a receptor on the surface of the host cell. Without the B subunits, the A subunit has no way of attaching to or entering the cell, and thus no way to exert its toxic effect. Cholera toxin, shiga toxin, and SubAB toxin all have B subunits that are made up of five identical protein components, meaning that their B subunits are homopentamers. Pertussis toxin is different where its pentameric ring is made up of four different protein components, where one of the components is repeated to form a heteropentamer.

Mechanisms

Cholera toxin, pertussis toxin, and shiga toxin all have their targets in the cytosol of the cell. After their B subunit binds to receptors on the cell surface, the toxin is enveloped by the cell and transported inside either through clathrin-dependent endocytosis or clathrin-independent endocytosis.

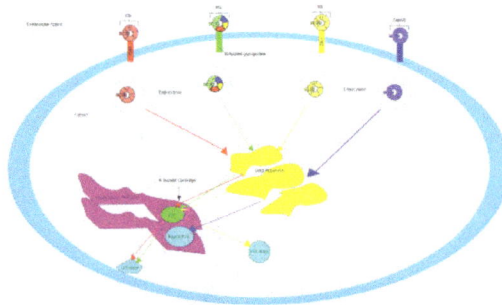

The mechanism pathways for the four AB5 toxins: cholera toxin, pertussis toxin, shiga toxin, and subtilase cytotoxin.

For the cholera toxin, the principal glycolipid receptor for the cholera toxin is ganglioside GM1. After endocytosis to the golgi apparatus, the toxin is redirected to the endoplasmic reticulum. In order for the A subunit to reach its target, a disulfide bond between the A1 and A2 domain must be broken. This breakage is catalyzed by a protein disulfide isomerase that is in the endoplasmic reticulum. Following separation, the A1 domain unfolds and is redirected back to the cytosol where it refolds and catalyzes ADP-ribosylation of certain G protein alpha subunits. In doing so, the downstream effects of the G protein signal transduction pathway is disrupted by activating adenylate cyclase. This causes a higher concentration of cAMP in the cell, which disrupts the regulation of ion transport mechanisms.

The pertussis toxin does not have a specific receptor, and binds to sialylated glycoproteins. After endocytosis, pertussis toxin's mechanism is the same as cholera toxin.

The main receptor for the shiga toxin is globotriaosylceramide or Gb3. Shiga toxin is also brought to the golgi apparatus before being directed to the endoplasmic reticulum for PDI to cleave the disulfide bond. Shiga toxin's A subunit is then brought back into the cytosol and inhibits eukaryotic protein synthesis with its RNA N-glycosidase activity by cleaving a specific adenine base on 28S ribosomal RNA that will ultimately cause cell death.

SubAB's target is in the endoplasmic reticulum of the cell and is brought into the cell through clathrin-mediated endocytosis. The glycan receptor for SubAB usually ends with an α2-3-linked N-Glycolylneuraminic acid (Neu5Gc). SubAB has an A subunit where it acts as a serine protease and cleaves Bip/GRP78, an endoplasmic reticulum chaperone. The cleavage of this chaperone causes cellular stress through protein inhibition, and consequently death of the cell.

Medical Uses

Cancer Treatment

B subunits of the AB5 toxins have the affinity towards binding glycan which some type of tumors seem to possess making it an easy target. One example is that of StxB which specifically binds with CD77 which shows expression on the surface of cancerous cells such as colon, pancreas, breast, and many more. Once StxB targets a cancerous cell, it delivers the A subunit of the toxin which eventually kills the cancerous cell.

Yet another method is by using ER stress-inducing drugs which have been tested in mice to show positive synergistic responses. This is accomplished through fusion of epidermal growth factor(EGF) with SubAB's A subunit. Cancer cells that express receptors for EGF will then experience SubAB toxicity.

Vaccines

Another use of AB5 toxins is using members of the LT family as adjuvants. This allows the toxin to promote immunological responses such as IgG2a, IgA, and Th17 to fight for instance gastric *Helicobacter pylori* infection when a vaccine is given.

In addition to some of these AB5 toxins being used to create vaccines to prevent bacterial infection, they are also being researched to work as a conjugate to prevent viral infections. For example, systemic immunization along with co-administered intra-nasal delivery of virus-cholera toxin conjugate vaccine induced a virus-specific antibody response and showed some degree of protection to the upper respiratory tract from Sendai virus.

Recent Areas of Research

New advancements in biotechnological experimental methods such as the use of Bessel beam plane illumination microscopy and FRET-based sensor molecules can better demonstrate dynamic structures of gap junction plaques. For these experiments, different types of AB5 toxins can be used to induce the fast formation of tCDR in E.Coli cells. The response can then be recorded using cAMP concentration fluctuations in gap junction-coupled cells using FRET-based sensor constructs. Research suggests that CDRs could perhaps be linked with rapid rearrangement of lipids and protein in connexin channels within the gap junction plaques. This can further help us understand the signaling cascade that follows a cellular loss of K+ when exposed to bacterial infection.

The SubAB toxin has been seen to demonstrate specificity to a binding protein, BiP. This characteristic has been utilized to study the role of the cellular BiP itself, along with Endoplasmic-reticulum-associated degradation in stressed HeLa cells.

Actin Assembly-inducing Protein

The Actin assembly-inducing protein (ActA) is a protein encoded and used by *Listeria monocytogenes* to propel itself through a mammalian host cell. ActA is a bacterial surface protein comprising a membrane-spanning region. In a mammalian cell the bacterial ActA interacts with the Arp2/3 complex and actin monomers to induce actin polymerization on the bacterial surface generating an actin comet tail. The gene encoding ActA is named *actA* or prtB.

Introduction

As soon as *L. monocytogenes* bacteria are ingested by humans, they get internalized into intestinal epithelium cells and rapidly try to escape their internalization vacuole. In the cytosol they start to polymerize actin on their surface by the help of the ActA protein. It has been shown that ActA is not only necessary but also sufficient to induce motility of bacteria in the absence of other bacterial factors.

Discovery of ActA

ActA was discovered by analysing lecithinase-negative Tn*917-lac* Listeria mutants because of the phenotype that they were unable to spread from cell to cell. These mutant bacteria still escaped from the phagosomes as efficiently as wild-type bacteria and multiplied within the infected cells but they were not surrounded by actin like wild-type bacteria. Further analysis showed, that Tn917-lac had inserted into *actA*, the second gene of an operon. The third gene of this operon, *plcB*, encodes the *L. monocytogenes* lecithinase. To determine whether *actA* itself, *plcB* or other co-transcribed downstream regions are involved in actin assembly, mutations in the appropriate genes were generated. All mutants except the *actA* mutants were similar to wild-type concerning association with F-actin and cell-cell spreading. Complementation with *actA* restored wild-type phenotype in the *actA* mutants.

Function

ActA is a protein which acts as a mimic of Wiskott-Aldrich syndrome protein (WASP), a nucleation promoting factor (NPF) present in host cells. NPFs in the mammalian cell recruit and bind to the already existing actin-related-protein 2 and 3 complex (Arp2/3 complex) and induce an activating conformational change of the Arp2/3 complex. Due to this conformational change, NPFs initiate polymerization of a new actin filament at a 70° angle, which leads to the characteristic Y-branched actin structures in the leading edge of motile cells. ActA localizes to the old pole of the bacterium and spans both the bacterial cell membrane and the cell wall, lateral diffusion is inhibited; thus ActA localizes in a polarized and anchored manner on the bacterial surface. Consequently actin polymerization only starts in this region on the surface of the bacterium. Expression of ActA is induced only after entering a mammalian host cell.

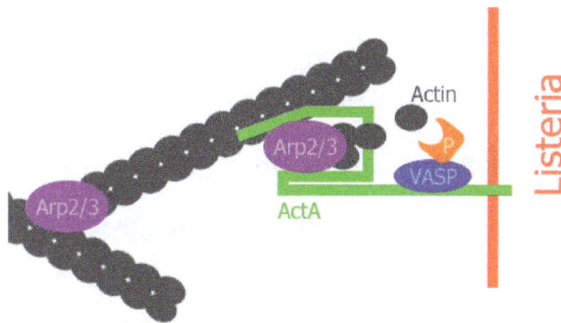

Fig. 1 Actin assembly induced by bacterial protein ActA (shown in green). Mammalian proteins involved in this process are: Profilin (P), Vasodilator-stimulated phosphoprotein (VASP) and actin-related-protein 2 and 3 complex (Arp2/3 complex) as well as actin.

Actin filament assembly generates the force that pushes the bacterium in the mammalian host cytoplasm forward. Continuous actin polymerization is sufficient for motility in the cytoplasm and even for infection of adjacent cells.

Research

New data indicates that ActA plays a role also in vacuolar disruption. A deletion mutant of ActA was defective in permeabilizing the vacuole. An 11 amino acid stretch of the N-terminus of the acidic region (32-42) was shown to be important for disruption of the phagosome.

Structure

The primary proteinous product of the *actA* gene consists of 639 amino acids and includes the signal peptide (first N-terminal 29 amino acids) and the ActA chain (C-terminal 610 amino acids). Therefore the sequence of the mature ActA protein consist of 610 amino acids. ActA has a molecular weight of 70,349 Da and is a surface protein.

ActA is a natively unfolded protein which can be divided into three functional domains (Fig. 2):

- N-terminal domain that is highly charged: amino acid residues 1-234

- central domain with proline-rich repeats: amino acid residues 235-394

- C-terminal domain with a transmembrane domain: amino acid residues 395-610

N-terminal Domain

The first 156 amino acids of the N-terminal domain consist of three regions (Fig. 2):

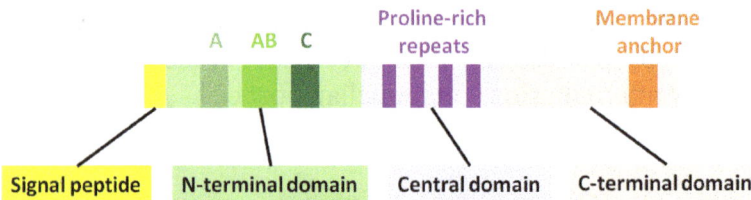

Fig. 2 The ActA protein and its functional domains

- A-region with a stretch of acidic residues: 32-45

- AB-region, an actin monomer-binding region: 59-102

- C-region, a cofilin homology sequence: 145-156

The N-terminal portion of ActA plays an important role in actin polymerization. The domain displays consensus elements present in eukaryotic WASP family NPFs which include an actin monomer-binding region as well as an Arp2/3 binding C (central or cofilin homology) and A (acidic) region. The actin monomer-binding region of ActA has functional properties like the WASP-Homology-2 (WH2) or V domain, but differs in the sequence. Thus in WASP-family NPFs the order of the domains is WH2 followed by C,and then by A, which is not the case in ActA.

Central Domain

The central proline-rich region of ActA is crucial for ensuring efficient bacterial motility. There are four proline-rich repeats containing either FPPPP or FPPIP motifs. These regions mimic those of the host cell cytoskeletal protein zyxin, vinculin and palladin, known to associate with focal adhesions or stress fibers. The vasodilator-stimulated phosphoprotein (VASP) can bind through its Ena/VASP homology 1 domain (EVH1 domain) to the central proline-rich region and recruits profilin, an actin monomer binding protein, which itself promotes polymerization at barbed ends of actin filaments. Furthermore, VASP seems to interact with F-actin through its carboxy-terminal EVH2 domain, which provides a linkage of the bacterium to the tail. This statement is supported by the fact that ActA can bind multiple Ena/VASP proteins simultaneously and has a high affinity between ActA and Ena/VASP. VASP has been shown to reduce the frequency actin-Y-branches in vitro and thus increases the proportion of filaments which are organized in a parallel alignment in comet tails.

C-terminal Domain

The C-terminal domain of ActA has a hydrophobic region which anchors the protein in the bacterial membrane.

In summary, besides

- the absence of sequence homology in the actin-binding-region and

- an alteration in the sequence of ARP2/3 activating domains typical for WASP-family NPFs (V(WH2)-C-A),

- a major difference between ActA and host NPFs is that ActA does not have elements that bind to regulatory proteins such as Rho family GTPases. This structural difference between ActA and host NPFs can be advantageous for *L. monocytogenes* and its pathogenesis because the actin nucleation activity of *L. monocytogenes* is independent of host regulation.

Analogues

WASP/N-WASP, which is functionally mimicked by ActA, is highly conserved in eukaryotes. It is an important actin-cytoskeleton organizer and is critical for processes such as endocytosis and cell motility. Activated by Cdc42, a Rho-family small GTPase, WASP/N-WASP activates the Arp2/3 complex, which leads to rapid actin polymerization.

Actin-based Motility of Other Pathogens

In *Shigella* the protein IcsA activates N-WASP, which in non-infected mammalian cells is activated by the GTPase Cdc42. Active N-WASP/WASP leads to actin polymerization by activating the Arp2/3 complex. In contrast, the *Listeria* ActA protein interacts with and activates directly the Arp2/3 complex.

The *Rickettsia* RickA protein is also able to activate the Arp2/3 complex in a WASP-like manner. In contrast to *Listeria*, the actin filaments are organized in long, unbranched parallel bundles. The Arp2/3 complex is only localized near the bacterial surface and thus it is assumed that a more frequent Arp2/3 complex-independent elongation occurs.

In *Burkholderia pseudomallei* BimA initiates actin polymerization in vitro. It is assumed that intracellular migration of this bacterium functions independently of the Arp2/3 complex.

Bacterial Effector Protein

Bacterial effectors are proteins secreted by pathogenic bacteria into the cells of their host, usually using a type 3 secretion system (TTSS/T3SS) or a type 4 secretion system (TFSS/T4SS). Some bacteria inject only a few effectors into their host's cells while others may inject dozens or even hundreds. Effector proteins may have many different activities, but usually help the pathogen to invade host tissue, suppress its immune system, or otherwise help the pathogen to survive. Effector proteins are usually critical for virulence. For instance, in the causative agent of plague (*Yersinia pestis*), the loss of the T3SS is sufficient to render the bacteria completely avirulent, even when they are directly introduced into the bloodstream. Gram negative microbes are also suspected to deploy bacterial outer membrane vesicles to translocate effector proteins and virulence factors via a novel membrane vesicle trafficking secretory pathway, in order to modify their environment or attack/invade target cells, for example, at the host-pathogen interface.

Diversity of Bacterial Effectors

Many pathogenic bacteria are known to have secreted effectors but for most species the exact number is unknown. Once a pathogen genome has been sequenced, effectors can be predicted based on protein sequence similarity, but such predictions are not always precise. More importantly, it is difficult to prove experimentally that a predicted effector is actually secreted into a host cell because the amount of each effector protein is tiny. For instance, Tobe et al. (2006) predicted more than 60 effectors for pathogenic *E. coli* but could only show for 39 that they are secreted into human Caco-2 cells. Finally, even within the same bacterial species, different strains often have different repertoires of effectors. For instance, the plant pathogen *Pseudomonas syringae* has 14 effectors in one strain, but more than 150 have been found in multiple different strains.

Species	number of effectors
Chlamydia (multiple species)	16+
E. coli EHEC (O157:H7)	40-60
E. coli (EPEC)	>20
Legionella pneumophila	>330 (T4SS)

Pseudomonas aeruginosa	4
Pseudomonas syringae	14 (>150 in multiple strains)
Salmonella enterica	60+
Yersinia (multiple species)	14

Mechanism of Action

Given the diversity of effectors, they affect a wide variety of intracellular processes. The T3SS effectors of pathogenic *E. coli, Shigella, Salmonella*, and *Yersinia* regulate actin dynamics to facilitate their own attachment or invasion, subvert endocytic trafficking, block phagocytosis, modulate apoptotic pathways, and manipulate innate immunity as well as host responses.

Phagocytosis. Phagocytes are immune cells that can recognize and "eat" bacteria. Phagocytes recognize bacteria directly [e.g., through the so-called scavenger receptor A which recognizes bacterial lipopolysaccharide (LPS)] or indirectly through antibodies (IgG) and complement proteins (C3bi) which coat the bacteria and are recognized by the Fcγ receptors and integrin$\alpha_m\beta_2$ (complement receptor 3). For instance, intracellular *Salmonella* and *Shigella* escape phagocytic killing through manipulation of endolysosomal trafficking. *Yersinia* predominantly survives extracellularly using the translocation of effectors to inhibit cytoskeletal rearrangements and thus phagocytosis. EPEC/EHEC inhibit both transcytosis through M cells and internalization by phagocytes. *Yersinia* inhibits phagocytosis through the concerted actions of several effector proteins, including YopE which acts as a RhoGAP and inhibits Rac-dependent actin polymerization.

Endocytic trafficking. Several bacteria, including *Salmonella* and *Shigella*, enter the cell and survive intracellularly by manipulating the endocytic pathway. Once internalized by host cells *Salmonella* subverts the endolysosome trafficking pathway to create a *Salmonella*-containing vacuole (SCV), which is essential for its intracellular survival. As the SCVs mature they travel to the microtubule organizing center (MTOC), a perinuclear region adjacent to the Golgi, where they produce *Salmonella* induced filaments (Sifs) dependent on the T3SS effectors SseF and SseG. By contrast, internalized *Shigella* avoids the endolysosome system by rapidly lysing its vacuole through the action of the T3SS effectors IpaB and C although the details of this process are poorly understood.

Secretory pathway. Some pathogens, such as EPEC/EHEC disrupt the secretory pathway. For instance, their effector EspG can reduce the secretion of interleukin-8 (IL-8), and thus affect the immune system (immunomodulation). EspG functions as a Rab GTPase-activating protein (Rab-GAP), trapping Rab-GTPases in their inactive GDP bound form, and reducing ER–Golgi transport (of IL-8 and other proteins).

Apoptosis (programmed cell death). Apoptosis is usually a mechanism of defense to infection, given that apoptotic cells eventually attract immune cells to remove them and the pathogen. Many pathogenic bacteria have developed mechanisms to prevent apoptosis, not the least to maintain their host environment. For instance, the EPEC/EHEC effectors NleH and NleF block apoptosis. Similarly, the *Shigella* effectors IpgD and OspG (a homolog of NleH) block apoptosis, the former by phosphorylating and stabilizing the double minute 2 protein (MDM2) which in turn leads to a block of NF-kB-induced apoptosis. *Salmonella* inhibits apoptosis and activates pro-survival signals, dependent on the effectors AvrA and SopB, respectively.

Induction of cell death. In contrast to inhibition of apoptosis, several effectors appear to induce programmed cell death. For instance, EHEC effectors EspF, EspH, and Cif induce apoptosis.

Inflammatory response. Human cells have receptors that recognize pathogen-associated molecular patterns (PAMPs). When bacteria bind to these receptors, they activate signaling cascades such as the NF-kB and MAPK pathways. This leads to expression of cytokines, immunomodulating agents, such as interleukins and interferons which regulate immune response to infection and inflammation. Several bacterial effectors affect NF-kB signaling. For instance, the EPEC/EHEC effectors NleE, NleB, NleC, NleH, and Tir are immunosuppressing effectors that target proteins in the NF-kB signaling pathway. NleC has been shown to cleave the NF-kB p65 subunit (RelA), blocking the production of IL-8 following infection. NleH1, but not NleH2, blocks translocation of NF-kB into the nucleus. The Tir effector protein inhibits cytokine production. Similarly, YopE, YopP, and YopJ (in *Yersinia enterocolitica*, *Yersinia pestis*, and *Yersinia pseudotuberculosis* respectively) target the NF-kB pathway. YopE inhibits activation of NF-kB, which in part prevents the production of IL-8.

Databases and Online Resources

- Effectors.org – A database of predicted bacterial effectors. Includes an interactive server to predict effectors.

- Bacterial Effector Proteins and their domains/motifs (from Paul Dean's lab)

- T3DB – A database of Type 3 Secretion System (T3SS) proteins

- T3SE – T3SS Database

- BEAN 2.0: an integrated web resource for the identification and functional analysis of type III secreted effectors

cAMP Receptor Protein

cAMP receptor protein (CRP; also known as catabolite activator protein, CAP) is a regulatory protein in bacteria. CRP protein binds cAMP, which causes a conformational change that allows CRP to bind tightly to a specific DNA site in the promoters of the genes it controls. CRP then activates transcription through direct protein–protein interactions with RNA polymerase.

The genes regulated by CRP are mostly involved in energy metabolism, such as galactose, citrate, or the PEP group translocation system. In *Escherichia coli*, cyclic AMP receptor protein (CRP) can regulate the transcription of more than 100 genes.

The signal to activate CRP is the binding of cyclic AMP. Binding of cAMP to CRP leads to a long-distance signal transduction from the N-terminal cAMP-binding domain to the C-terminal domain of the protein, which is responsible for interaction with specific sequences of DNA.

At "Class I" CRP-dependent promoters, CRP binds to a DNA site located upstream of core promoter elements and activates transcription through protein–protein interactions between "activating region 1" of CRP and the C-terminal domain of RNA polymerase alpha subunit. At "Class II" CRP-dependent promoters, CRP binds to a DNA site that overlaps the promoter -35 element and activates transcription through two sets of protein–protein interactions: (1) an interaction between "activating region 1" of CRP and the C-terminal domain of RNA polymerase alpha subunit, and (2) an interaction between "activating region 2" of CRP and the N-terminal domain of RNA polymerase alpha subunit. At "Class III" CRP-dependent promoters, CRP functions together with one or more "co-activator" proteins.

At most CRP-dependent promoters, CRP activates transcription primarily or exclusively through a "recruitment" mechanism, in which protein–protein interactions between CRP and RNA polymerase assist binding of RNA polymerase to the promoter.

Cholera Toxin

Cholera toxin (also known as choleragen and sometimes abbreviated to CTX, Ctx or CT) is protein complex secreted by the bacterium *Vibrio cholerae*. CTX is responsible for the massive, watery diarrhea characteristic of cholera infection.

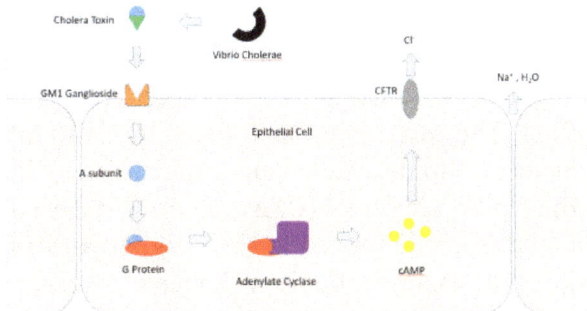

Cholera toxin mechanism

History

Cholera toxin was discovered in 1959 by Late Prof.S.N.De (Ref.1) at Kolkata (India)

Structure

The cholera toxin is an oligomeric complex made up of six protein subunits: a single copy of the A subunit (part A, enzymatic), and five copies of the B subunit (part B, receptor binding), denoted as AB_5. Subunit B binds while subunit A activates the G protein which activates adenylate cyclase. The three-dimensional structure of the toxin was determined using X-ray crystallography by Zhang *et al.* in 1995.

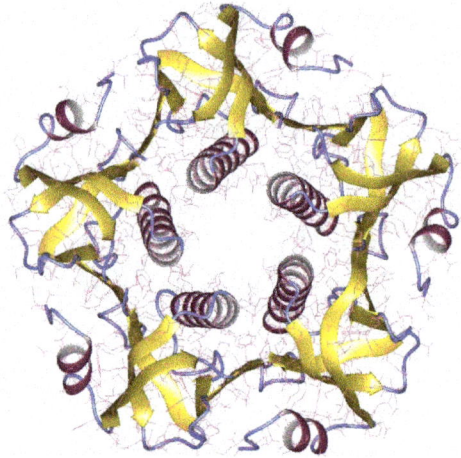

Cholera toxin B pentamer, Vibrio cholerae.

The five B subunits—each weighing 11 kDa, form a five-membered ring. The A subunit which is 28 kDa, has two important segments. The A1 portion of the chain (CTA1) is a globular enzyme payload that ADP-ribosylates G proteins, while the A2 chain (CTA2) forms an extended alpha helix which sits snugly in the central pore of the B subunit ring.

This structure is similar in shape, mechanism, and sequence to the heat-labile entero-toxin secreted by some strains of the *Escherichia coli* bacterium.

Pathogenesis

Cholera toxin acts by the following mechanism: First, the B subunit ring of the cholera toxin binds to GM1 gangliosides on the surface of target cells. The B subunit can also bind to cells lacking GM1. The toxin then most likely binds to other types of glycans attached to proteins instead of lipids. Once bound, the entire toxin complex is endocy-tosed by the cell and the cholera toxin A1 (CTA1) chain is released by the reduction of a disulfide bridge. The endosome is moved to the Golgi apparatus, where the A1 pro-tein is recognized by the endoplasmic reticulum chaperon, protein disulfide isomerase. The A1 chain is then unfolded and delivered to the membrane, where Ero1 triggers the release of the A1 protein by oxidation of protein disulfide isomerase complex. As the

A1 protein moves from the ER into the cytoplasm by the Sec61 channel, it refolds and avoids deactivation as a result of ubiquitination.

CTA1 is then free to bind with a human partner protein called ADP-ribosylation factor 6 (Arf6); binding to Arf6 drives a change in the shape of CTA1 which exposes its active site and enables its catalytic activity. The CTA1 fragment catalyses ADP-ribosylation of the Gs alpha subunit ($G\alpha_s$) proteins using NAD. The ADP-ribosylation causes the $G\alpha_s$ subunit to lose its catalytic activity of inactivating GTP by hydrolyzing it to GDP + P_i, effectively increasing GTP concentration as less is being converted back to inactive GDP. Increased $G\alpha_s$ activation leads to increased adenylate cyclase activity, which increases the intracellular concentration of 3',5'-cyclic AMP (cAMP) to more than 100-fold over normal and over-activates cytosolic PKA. These active PKA then phosphorylate the cystic fibrosis transmembrane conductance regulator (CFTR) chloride channel proteins, which leads to ATP-mediated efflux of chloride ions and leads to secretion of H_2O, Na^+, K^+, and HCO_3^- into the intestinal lumen. In addition, the entry of Na^+ and consequently the entry of water into enterocytes are diminished. The combined effects result in rapid fluid loss from the intestine, up to 2 liters per hour, leading to severe dehydration and other factors associated with cholera, including a rice-water stool.

The pertussis toxin (also an AB_5 protein) produced by *Bordetella pertussis* acts in a similar manner with the exception that it ADP-ribosylates the $G\alpha_i$ subunit, rendering it inactive and unable to inhibit adenylyl cyclase production of cAMP (leading to constitutive production).

Expression of cholera toxin was shown to be prevented by the drug virstatin, which acts to prevent its transcription. It has been identified as a possible treatment for cholera.

Origin

The gene encoding the cholera toxin is introduced into *V. cholerae* by horizontal gene transfer. Virulent strains of *V. cholerae* carry a variant of temperate bacteriophage called the CTXf or CTXφ Bacteriophage.

Applications

Because the B subunit appears to be relatively non-toxic, researchers have found a number of applications for it in cell and molecular biology. It is routinely used as a neuronal tracer.

Treatment of cultured rodent neural stem cells with cholera toxin induces changes in the localization of the transcription factor Hes3 and increases their numbers.

GM1 gangliosides are found in lipid rafts on the cell surface. B subunit complexes labelled with fluorescent tags or subsequently targeted with antibodies can be used to identify rafts.

Crescentin

Crescentin is a protein which is a bacterial relative of the intermediate filaments found in eukaryotic cells. Just as tubulins and actins, the other major cytoskeletal proteins, have prokaryotic homologs in, respectively, the FtsZ and MreB proteins, intermediate filaments are linked to the crescentin protein.

Role in Cell Shape

Crescentin was recently discovered in the prokaryote *Caulobacter crescentus*, an aquatic bacterium which uses its crescent-shaped cells for enhanced motility. The crescentin protein is located on the concave face of these cells and appears to be necessary for their shape, as mutants lacking the protein form rod-shaped cells. To influence the shape of the *Caulobacter* cells, the helices of crescentin filaments associate with the cytoplasmic side of the cell membrane on one lateral side of the cell. This induces a curved cell shape in younger cells, which are shorter than the helical pitch of crescentin, but induces a spiral shape in older, longer cells.

Protein Structure

Like eukaryotic intermediate filaments, crescentin organizes into filaments and is present in a helical structure in the cell. Crescentin is necessary for both shapes of the *Caulobacter* prokaryote (vibroid/crescent-shape and helical shape, which it may adopt after a long stationary phase). The crescentin protein has 430 residues; its sequence mostly consists of a pattern of 7 repeated residues which form a coiled-coil structure. The DNA sequence of the protein has sections very similar to the eukaryotic keratin and lamin proteins, mostly involving the coiled-coil structure. Researchers Ausmees et al. recently proved that, like animal intermediate filament proteins, crescentin has a central rod made up of four coiled-coil segments. Both intermediate filament and crescentin proteins have a primary sequence including four α-helical segments along with non-α-helical linker domains. An important difference between crescentin and animal intermediate filament proteins is that crescentin lacks certain consensus sequence elements at the ends of the rod domain which are conserved in animal lamin and keratin proteins.

Assembly of Filaments

Eukaryotic intermediate filament proteins assemble into filaments of 8-15 nm within the cell without the need for energy input, that is, no need for ATP or GTP. Ausmees et al. continued their crescentin research by testing whether the protein could assemble into filaments in this manner *in vitro*. They found that crescentin proteins were indeed able to form filaments about 10 nm wide, and that some of these filaments organized laterally into bundles, just as eukaryotic intermediate filaments do. The similarity of crescentin protein to intermediate filament proteins suggests an evolutionary linkage between these two cytoskeletal proteins.

Iron-starvation-induced Protein A

IsiA is a photosynthesis-related chlorophyll-containing protein found in cyanobacteria. It belongs to the chlorophyll-a/b-binding family of proteins, and has been shown to have a photoprotection role in preventing oxidative damage via energy dissipation. It was originally identified under Fe-starvation, and thus received the name Iron-Starvation-Induced protein A. However, the protein has more recently been found to respond to a variety of stress conditions such as high irradiance. It can aggregate with carotenoids and form rings around the PSI reaction center complexes to aid in photoprotective energy dissipation.

Antenna Function

IsiA functions as an antenna for photosystem I (PSI) under iron-limiting conditions, when phycobilisomes disappear. In the (PSI)3(Isi3)18 complex most of the harvested energy is probably used by PSI; in other PSI-containing supercomplexes a large part of the energy will probably not be used for light harvesting, but rather is dissipated to protect the organism from light damage. Under iron-starvation, it forms a complex with PSI trimers, where the trimer is surrounded by a ring composed of 18 IsiA subunits. When the PSI subunits PsaF and PsaJ are missing the ring is composed of 17 IsiA subunits, indicating that each IsiA subunit has a different interaction with the trimer. This suggests that the size of the PSI complex determines the number of IsiA units in the surrounding ring. In the absence of PsaL, it has a tendency to form incomplete rings with PSI monomers, suggesting PsaL helps form the rings. Also it can form other aggregates of varying sizes depending on the level of iron-deprivation.

Photoprotective Function

IsiA aggregates, forming empty multimeric rings (without PSI) also accumulate and are very abundant in long-term iron-depleted cells. When isolated, these aggregates are in a strongly quenched state, suggesting they are responsible for thermal dissipation of absorbed energy. IsiA is also synthesized in cells grown under high light irradiances, protecting from photodamage

Lac Repressor

The *lac* repressor is a DNA-binding protein which inhibits the expression of genes coding for proteins involved in the metabolism of lactose in bacteria. These genes are repressed when lactose is not available to the cell, ensuring that the bacterium only invests energy in the production of machinery necessary for uptake and utilization of lactose when lactose is present. When lactose becomes available, it is converted into allolactose, which inhibits the *lac* repressor's DNA binding ability. Loss of DNA binding by the *lac* repressor is required for transcriptional activation of the operon.

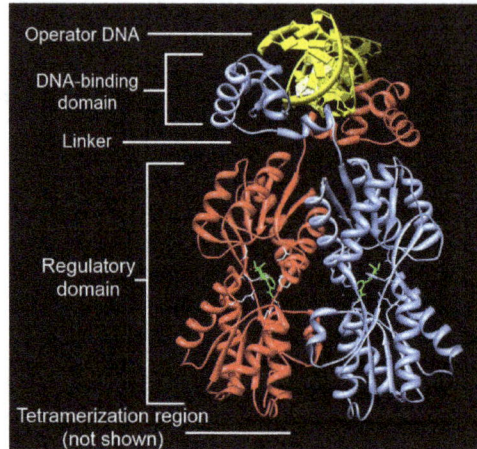

Annotated crystal structure of dimeric LacI. Two monomers (of four total) co-operate to bind each DNA operator sequence. Monomers (red and blue) contain DNA binding and core domains (labeled) which are connected by a linker (labeled). The C-terminal tetramerization helix is not shown. The repressor is shown in complex with operator DNA (gold) and ONPF (green), an anti-inducer ligand (*i.e.* a stabilizer of DNA binding)

Function

The *lac* repressor (LacI) operates by a helix-turn-helix motif in its DNA binding domain binding base-specifically to the major groove of the operator region of the *lac* operon, with base contacts also made by residues of symmetry-related alpha helices, the "hinge" helices, which bind deeply in the minor groove. This DNA binding causes the specific affinity of RNA polymerase for the promoter sequence to increase sufficiently that it cannot escape the promoter region and enter elongation, and so prevents transcription of the mRNA coding for the Lac proteins. When lactose is present, allolactose binds to the *lac* repressor, causing an allosteric change in its shape. In its changed state, the *lac* repressor is unable to bind tightly to its cognate operator. This effect is referred to as induction, because it *induces*, rather than represses, expression of the metabolic genes. *In vitro,* Isopropyl β-D-1-thiogalactopyranoside (IPTG) is a commonly used allolactose mimic which can be used to induce transcription of genes being regulated by *lac* repressor.

Structure

Structurally, the *lac* repressor protein is a homo-tetramer. The tetramer contains two DNA binding subunits composed of two monomers each (sometimes called "dimeric *lac* repressor"). These subunits dimerize to form a tetramer capable of binding two operator sequences. Each monomer consists of four distinct regions:

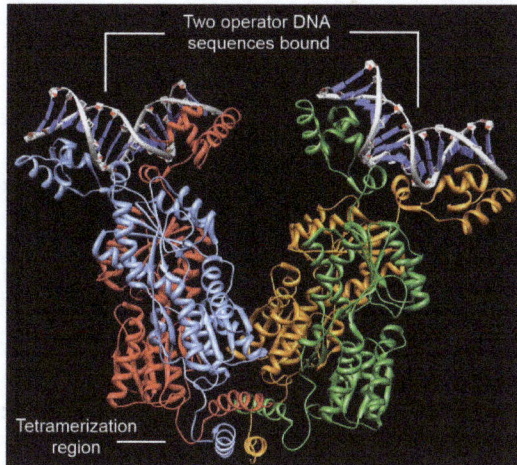

Tetrameric LacI binds two operator sequences and induces DNA looping. Two dimeric *LacI* functional subunits (red+blue and green+orange) each bind a DNA operator sequence (labeled). These two functional subunits are coupled at the tetramerization region (labeled); thus, tetrameric *LacI* binds two operator sequences. This allows tetrameric *LacI* to induce DNA looping.

- an N-terminal DNA-binding domain (in which two LacI proteins bind a single operator site)

- a regulatory domain (sometimes called the core domain, which binds allolactose, an allosteric effector molecule)

- a linker that connects the DNA-binding domain with the core domain (sometimes called the hinge helix, which is important for allosteric communication)

- a C-terminal tetramerization region (which joins four monomers in an alpha-helix bundle)

DNA binding occurs via an N-terminal helix-turn-helix structural motif and is targeted to one of several operator DNA sequences (known as O_1, O_2 and O_3). The O_1 operator sequence slightly overlaps with the promoter, which increases the affinity of RNA polymerase for the promoter sequence such that it cannot enter elongation and remains in Abortive initiation. Additionally, because each tetramer contains two DNA-binding subunits, binding of multiple operator sequences by a single tetramer induces DNA looping.

Discovery

The *lac* repressor was first isolated by Walter Gilbert and Benno Müller-Hill in 1966. They were able to show, *in vitro*, that the protein bound to DNA containing the *lac* operon, and released the DNA when IPTG was added. (IPTG is an allolactose analog.) They were also able to isolate the portion of DNA bound by the protein by using the enzyme deoxyribonuclease, which breaks down DNA. After treatment of the repressor-DNA complex, some DNA remained, suggesting that it had been masked by the repressor. This was later confirmed.

These experiments confirmed the mechanism of the *lac* operon, earlier proposed by Jacques Monod and Francois Jacob.

Methyl-accepting Chemotaxis Protein

Methyl-accepting chemotaxis protein (MCP) is a transmembrane sensor protein of bacteria. Use of the MCP allows bacteria to detect concentrations of molecules in the extracellular matrix so that the bacteria may smooth swim or tumble accordingly. If the bacteria detects rising levels of attractants (nutrients) or declining levels of repellents (toxins), the bacteria will continue swimming forward, or smooth swimming. If the bacteria detects declining levels of attractants or rising levels of repellents, the bacteria will tumble and re-orient itself in a new direction. In this manner, a bacterium may swim towards nutrients and away from toxins

Environmental diversity gives rise to diversity in bacterial signalling receptors, and consequently there are many genes encoding MCPs. For example, there are four well-characterised MCPs found in *Escherichia coli*: Tar (taxis towards aspartate and maltose, away from nickel and cobalt), Tsr (taxis towards serine, away from leucine, indole and weak acids), Trg (taxis towards galactose and ribose) and Tap (taxis towards dipeptides).

Structure

MCPs share similar structure and signalling mechanism. MCPs form dimers. Three dimers of MCP spontaneously form trimers. Trimers are complexed by CheA and CheW into hexagonal lattices. MCPs either bind ligands directly or interact with ligand-binding proteins, transducing the signal to downstream signalling proteins in the cytoplasm. Most MCPs contain: (a) an N-terminal signal peptide that is a transmembrane alpha-helix in the mature protein; (b) a poorly-conserved periplasmic receptor (ligand-binding) domain; (c) a transmembrane alpha-helix; (d) generally one or more HAMP domains and (e) a highly conserved C-terminal cytoplasmic domain that interacts with downstream signalling components. The C-terminal domain contains the methylated glutamate residues.

MCPs undergo two covalent modifications: deamidation and reversible methylation at a number of glutamate residues. Attractants increase the level of methylation, while repellents decrease it. The methyl groups are added by the methyl-transferase CheR and are removed by the methylesterase CheB.

Function

Binding a ligand causes a conformational change in the MCP receptor which translates down the hairpin structure. At the tip of the hairpin are two proteins that associate to the MCP: CheW and CheA. CheA acts as the sensor kinase. CheA has kinase activity and autophosphorylates itself on a histidyl residue when activated by the MCP. CheW is believed to be a transducer of the signal from the MCP to CheA. Activated CheA transfers its phosphoryl group to CheY, a response regulator. Phosphorylated CheY phosphorylates the basal body FliM which is connected to the flagellum. Phosphorylation of the basal body acts as a flagellar switch and changes the direction of rotation of the flagellum. This change in direction allows for alternation between smooth swimming and tumbling which biases the bacterial random walk towards attractant.

References

- Gutkind, edited by Toren Finkel, J. Silvio (2003). Signal Transduction and Human Disease. Hoboken, NJ: John Wiley & Sons. ISBN 0471448370.

- Welch, Matthew D. (2007). "Actin-based motility and cell-to-cell spread of Listeria monocytogenes". In Goldfine, Howard; Shen, Hao. Listeria monocytogenes: Pathogenesis and host response. New York: Springer. pp. 197–223. ISBN 978-0-387-49373-2.

- Faruque SM; Nair GB, eds. (2008). Vibrio cholerae: Genomics and Molecular Biology. Caister Academic Press. ISBN 978-1-904455-33-2 .

Various Bacterial Diseases

Bacteria can cause a number of diseases and some of these diseases are bacterial pneumonia, ehrlichia ruminantium, actinomycosis, leprosy and tuberculosis. Bacterial pneumonia is a particular type of pneumonia caused by bacteria and leprosy is also a disease caused by bacteria but the symptoms of this disease usually go unnoticed for initial couple of years. This chapter ha been carefully written to provide an easy understanding of the diseases caused by bacteria.

Bacterial Pneumonia

Bacterial pneumonia is a type of pneumonia caused by bacterial infection.

Signs and Symptoms

- Pneumonia

- Fever

- Rigors

- Cough

- Runny nose (either direct bacterial pneumonia or accompanied by primary viral pneumonia)

- Dyspnea - shortness of breath

- Chest pain

- Shaking chills

- Pneumococcal pneumonia can cause coughing up of blood, or hemoptysis, characteristically associated with "rusty" sputum

Types

Gram-positive

Streptococcus pneumoniae (J13) is the most common bacterial cause of pneumonia in all age groups except newborn infants. *Streptococcus pneumoniae* is a Gram-positive bacterium that often lives in the throat of people who do not have pneumonia.

Other important Gram-positive causes of pneumonia are *Staphylococcus aureus* (J15.2) and *Bacillus anthracis*.

Gram-negative

Gram-negative bacteria are seen less frequently: *Haemophilus influenzae* (J14), *Klebsiella pneumoniae* (J15.0), *Escherichia coli* (J15.5), *Pseudomonas aeruginosa* (J15.1), *Bordetella pertussis*, and *Moraxella catarrhalis* are the most common.

These bacteria often live in the gut and enter the lungs when contents of the gut (such as vomit or faeces) are inhaled.

Atypical

"Atypical" bacteria are *Coxiella burnetii*, *Chlamydophila pneumoniae* (J16.0), *Mycoplasma pneumoniae* (J15.7), and *Legionella pneumophila*.

Many people falsely believe they are called "atypical" because they are uncommon and/ or do not respond to common antibiotics and/or cause atypical symptoms. In reality, they are "atypical" because they do not gram stain as well as gram-negative and gram-positive organisms.

Pneumonia caused by *Yersinia pestis* is usually called pneumonic plague.

Pathophysiology

Bacteria typically enter the lung with inhalation, though they can reach the lung through the bloodstream if other parts of the body are infected. Often, bacteria live in parts of the upper respiratory tract and are continuously being inhaled into the alveoli. Once inside the alveoli, bacteria travel into the spaces between the cells and also between adjacent alveoli through connecting pores. This invasion triggers the immune system to respond by sending white blood cells responsible for attacking microorganisms (neutrophils) to the lungs. The neutrophils engulf and kill the offending organisms but also release cytokines that result in a general activation of the immune system. This results in the fever, chills, and fatigue common in bacterial and fungal pneumonia. The neutrophils, bacteria, and fluid leaked from surrounding blood vessels fill the alveoli and result in impaired oxygen transportation.

Bacteria often travel from the lung into the blood stream (bacteremia) and can result in serious illness such as sepsis and eventually septic shock, in which there is low blood pressure leading to damage in multiple parts of the body including the brain, kidney, and heart. They can also travel to the area between the lungs and the chest wall, called the pleural cavity.

Treatment

Antibiotics are the treatment of choice for bacterial pneumonia and ventilation (oxygen supplement) as supportive therapy. The antibiotic choice depends on the nature of the pneumonia, the microorganisms most commonly causing pneumonia in the geographical region, and the immune status and underlying health of the individual. In the United Kingdom, amoxicillin is used as first-line therapy in the vast majority of patients acquiring pneumonia in the community, sometimes with added clarithromycin. In North America, where the "atypical" forms of community-acquired pneumonia are becoming more common, clarithromycin, azithromycin, or fluoroquinolones as single therapy, have displaced the amoxicillin as first-line therapy. Local patterns of antibiotic-resistance should always be considered when initiating pharmacotherapy. In hospitalized individuals or those with immune deficiencies, local guidelines determine the selection of antibiotics.

Gram-positive Organisms

Streptococcus pneumoniae — amoxicillin (or erythromycin in patients allergic to penicillin); cefuroxime and erythromycin in severe cases. *Staphylococcus aureus* — flucloxacillin (to counteract the organism's β-lactamase).

Gram-negative Organisms

- *Haemophilus influenzae* — doxycycline; 2nd generation Cephalosporins such as Cefaclor

- *Klebsiella pneumoniae*

- *Escherichia coli*

- *Pseudomonas aeruginosa* — ciprofloxacin

- *Moraxella catarrhalis*

Atypical Organisms

- *Chlamydophila pneumoniae* — doxycycline

- *Chlamydophila psittaci* — erythromycin

- *Mycoplasma pneumoniae* — erythromycin

- *Coxiella burnetti* — doxycycline

- *Legionella pneumophila* — erythromycin, with rifampicin sometimes added.

People who have difficulty breathing due to pneumonia may require extra oxygen. An extremely sick individual may require artificial ventilation and intensive care as life-saving measures while his or her immune system fights off the infectious cause with the help of antibiotics and other drugs.

Prevention

Prevention of bacterial pneumonia is by vaccination against streptococcus pneumoniae (pneumococcal polysaccharide vaccine for adults and pneumococcal conjugate vaccine for children), Haemophilus influenzae type B, meningococcus, bordetella pertussis, bacillus anthracis, and yersinia pestis.

Ehrlichia Ruminantium

Heartwater (also known as cowdriosis, nintas and ehrlichiosis) is a tick-borne rickettsial disease of domestic and wild ruminants. It is caused by *Ehrlichia ruminantium* (formerly *Cowdria ruminantium*) - an intracellular gram-negative coccal bacterium (also referred to as *Rickettsia ruminantium*). The disease is spread by bont ticks, which are members of the genus *Amblyomma*. Affected mammals include cattle, sheep, goats, antelope, and buffalo, but the disease has the biggest economic impact on cattle production in affected areas. The disease's name is derived from the fact that fluid can collect around the heart or in the lungs of infected animals.

The disease is common in sub-Saharan Africa and some of the West Indian islands. It was first identified in sheep in South Africa in the 1830s, and had reached the Caribbean by 1980. The ticks which carry the disease occur in Africa and the Caribbean, and feed on a wide variety of vertebrate hosts. In the Caribbean, at least, the cattle egret has been implicated in the spread of heartwater since it colonized the islands in the 1950s. Animals often acquire the disease when moved on to heartwater infected grazing.

Cowdriosis is notifiable to the World Organisation for Animal Health.

Clinical Signs

Clinical disease is more common in young animals and non-native breeds. The clinical signs of disease are caused by an increased vascular permeability and consequent oedema and hypovolaemia.

The symptoms include neurological signs such as tremors and head pressing, respiratory signs such as coughing and nasal discharge, and systemic signs such as fever and loss of appetite. Physical examination may reveal petechiae of the mucous membranes, tachycardia and muffled heart sounds. Cowdriosis can also cause reproductive and gastrointestinal disease. It is frequently fatal.

Diagnosis

On post-mortem examination, a light yellow transudate that coagulates on exposure to air is often found within the thorax, pericardium and abdomen. Most fatal cases will have the hydropericardium that gives the disease its common name. Pulmonary oedema and mucosal congestion are regularly seen along with frothy fluid in the airways and cut surfaces of the lungs.

To definitively diagnose the disease, *C. ruminantium* must be demonstrated either in preparations of the hippocampus under Giemsa staining or by histopathology of brain or kidney.

Treatment and Control

During the early stages of disease animals may be treated with sulfonamides and tetracyclines. In advanced disease, prognosis is poor.

Amblyomma hebraeum, a vector of Heartwater disease

Tetracyclines can also be used prophylactically when animals are introduced into an area endemic with cowdriosis. A live blood vaccine is available for protection of young stock, but animals may require treatment for the disease post-vaccination. Ectoparasiticides dips can be used to reduce exposure the animals exposure to bont ticks. In areas endemic for heartwater there is likely to be use of dips against other ticks of domestic animals, such as *Rhipicephalus (Boophilus)* and *Hyalomma* species and this will usually contribute to control of vectors of *Ehrlichia ruminantium*.

Actinomycosis

Actinomycosis is a rare infectious bacterial disease caused by *Actinomyces* species. About 70% of infections are due to either *Actinomyces israelii* or *A. gerencseriae*. Infection can also be caused by other Actinomyces species, as well as *Propionibacterium propionicus*, which presents similar symptoms. The condition is likely to be polymicrobial aerobic anaerobic infection.

Signs and Symptoms

The disease is characterised by the formation of painful abscesses in the mouth, lungs, or gastrointestinal tract. Actinomycosis abscesses grow larger as the disease progresses, often over months. In severe cases, they may penetrate the surrounding bone and muscle to the skin, where they break open and leak large amounts of pus, which often contains characteristic granules (sulfur granules) filled with progeny bacteria. These granules are named due to their appearance but are not actually composed of sulfur. Sometimes there is difficulty in making the correct diagnosis. In addition to microbiological examinations, magnetic resonance imaging and immunoassays may also be helpful.

Causes

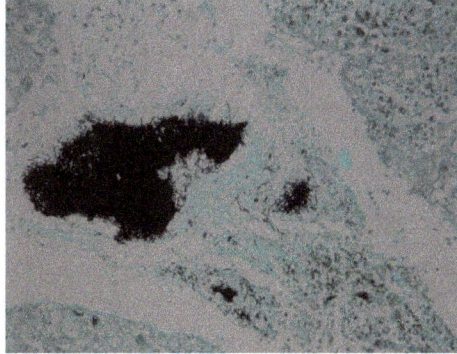

Actinomycosis Grocott's stain

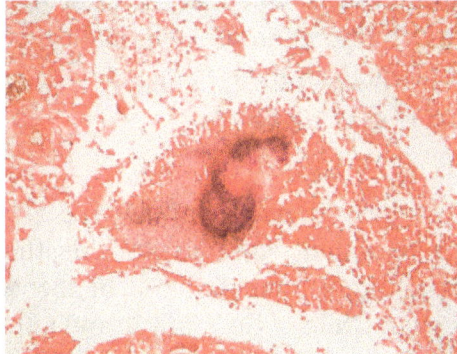

Actinomycosis Gram stain

Actinomycosis is primarily caused by any of several members of the bacterial genus *Actinomyces*. These bacteria are generally anaerobes. In animals, they normally live in the small spaces between the teeth and gums, causing infection only when they can multiply freely in anoxic environments. An affected human often has recently had dental work, poor oral hygiene, periodontal disease, radiation therapy, or trauma (broken jaw) causing local tissue damage to the oral mucosa, all of which predispose the person to developing actinomycosis. *Actinomyces israelii* is a normal commensal species part of the microbiota species of the lower reproductive tract of women. They are also normal commensals among the gut flora of the caecum; thus, abdominal actinomycosis can occur following removal of the appendix. The three most common sites of infection are decayed teeth, the lungs, and the intestines. It is important to note that actinomycosis does not occur in isolation from other bacteria. This infection depends on other bacteria (gram positive, gram negative, and cocci) to aid in invasion of tissue.

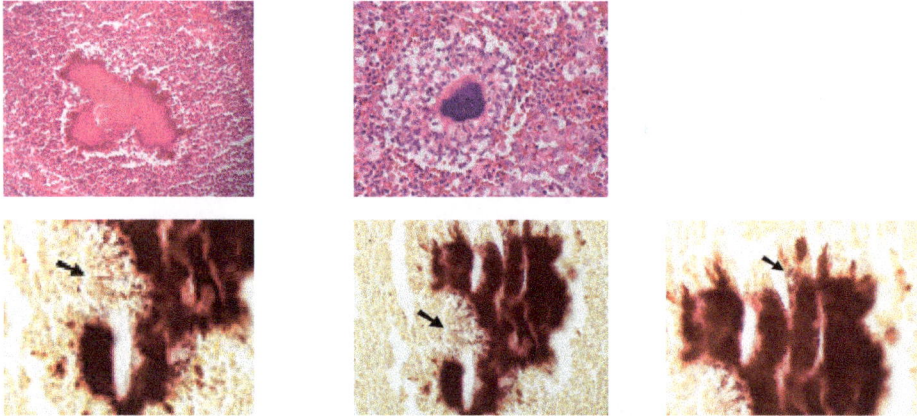

Treatment

Actinomyces bacteria are generally sensitive to penicillin, which is frequently used to treat actinomycosis. In cases of penicillin allergy, doxycyclin is used. Sulfonamides such as sulfamethoxazole may be used as an alternative regimen at a total daily dosage of 2-4 grams. Response to therapy is slow and may take months. Hyperbaric oxygen therapy may also be used as an adjunct to conventional therapy when the disease process is refractory to antibiotics and surgical treatment.

Epidemiology

There is a greater disease incidence in males between the ages of 20 and 60 years than in females. Before antibiotic treatments became available, the incidence in the Netherlands and Germany was 1 per 100,000 people/year. Incidence in the U.S. in the 1970s was 1 per 300,000 people/year, while in Germany in 1984, it was estimated to be 1 per 40,000 people/year. The use of intrauterine devices (IUDs) has increased incidence of genitourinary actinomycosis in females. Incidence of oral actinomycosis, which is harder to diagnose, has increased.

History

In 1877, pathologist Otto Bollinger described the presence of *Actinomyces bovis* in cattle, and shortly afterwards, James Israel discovered *Actinomyces israelii* in humans. In 1890, Eugen Bostroem isolated the causative organism from a culture of grain, grasses, and soil. After Bostroem's discovery there was a general misconception that actinomycosis was a mycosis that affected individuals who chewed grass or straw. The pathogen is still known as the "great masquerader". Bergey classified the organism as bacterial in 1939, but the disease remained classified as a fungus in the 1955 edition of the Control of Communicable Diseases in Man.

Violinist Joseph Joachim died of actinomycosis on 15 August 1907.

Other animals

Actinomycosis occurs rarely in humans but rather frequently in cattle as a disease called *lumpy jaw*. This name refers to the large abscesses that grow on the head and neck of the infected animal. It can also affect swine, horses, and dogs, and less often wild animals and sheep.

Leprosy

Leprosy, also known as Hansen's disease (HD), is a long-term infection caused by the bacilli *Mycobacterium leprae* and *Mycobacterium lepromatosis*. Initially, infections are without symptoms and typically remain this way for 5 to 20 years. Symptoms that develop include granulomas of the nerves, respiratory tract, skin, and eyes. This may result in a lack of ability to feel pain and thus loss of parts of extremities due to repeated injuries or infection due to unnoticed wounds. Weakness and poor eyesight may also be present.

Leprosy is spread between people. This is believed to occur through a cough or contact with fluid from the nose of an infected person. Leprosy occurs more commonly among those living in poverty. Contrary to popular belief, it is not highly contagious. The two main types of disease are based on the number of bacteria present: paucibacillary and multibacillary. The two types are differentiated by the number of poorly pigmented, numb skin patches present, with paucibacillary having five or fewer and multibacillary having more than five. The diagnosis is confirmed by finding acid-fast bacilli in a biopsy of the skin or by detecting the DNA using polymerase chain reaction.

Leprosy is curable with a treatment known as multidrug therapy. Treatment for paucibacillary leprosy is with the medications dapsone and rifampicin for six months. Treatment for multibacillary leprosy consists of rifampicin, dapsone, and clofazimine for 12 months. A number of other antibiotics may also be used. These treatments are provided free of charge by the World Health Organization. Globally in 2012, the number of chronic cases of leprosy was 189,000, down from some 5.2 million in the 1980s. The number of new cases was 230,000. Most new cases occur in 16 countries, with India accounting for more than half. In the past 20 years, 16 million people worldwide have been cured of leprosy. About 200 cases are reported per year in the United States.

Leprosy has affected humanity for thousands of years. The disease takes its name from the Latin word *lepra*, which means "scaly", while the term "Hansen's disease" is named after the physician Gerhard Armauer Hansen. Separating people by placing them in leper colonies still occurs in places such as India, China, and Africa. However, most colonies have closed since leprosy is not very contagious. Social stigma has been associated with leprosy for much of history, which continues to be a barrier to self-reporting and early treatment. Some consider the word "leper" offensive, preferring the phrase "person affected with leprosy". World Leprosy Day was started in 1954 to draw awareness to those affected by leprosy.

Signs and Symptoms

Leprosy is primarily a granulomatous disease of the peripheral nerves and mucosa of the upper respiratory tract; skin lesions (light or dark patches) are the primary external sign. If untreated, leprosy can progress and cause permanent damage to the skin, nerves, limbs, and eyes. Contrary to folklore, leprosy does not cause body parts to fall off, although they can become numb or diseased as a result of secondary infections; these occur as a result of the body's defenses being compromised by the primary disease. Secondary infections, in turn, can result in tissue loss, causing fingers and toes to become shortened and deformed, as cartilage is absorbed into the body.

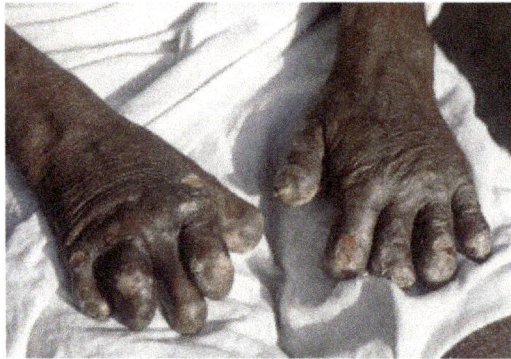
Hands deformed by leprosy

Leprosy in Tahiti, *circa* 1895

Cause

M. leprae

M. leprae and *M. lepromatosis* are the causative agents of leprosy. *M. lepromatosis* is a relatively newly identified mycobacterium isolated from a fatal case of diffuse lepromatous leprosy in 2008.

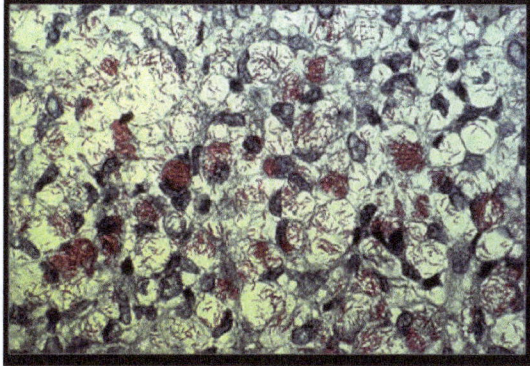

M. leprae, one of the causative agents of leprosy: As an acid-fast bacterium, *M. leprae* appears red when a Ziehl-Neelsen stain is used.

An intracellular, acid-fast bacterium, *M. leprae* is aerobic and rod-shaped, and is surrounded by the waxy cell membrane coating characteristic of the *Mycobacterium* genus.

Due to extensive loss of genes necessary for independent growth, *M. leprae* and *M. lepromatosis* are obligate intracellular pathogens, and unculturable in the laboratory, a factor that leads to difficulty in definitively identifying the organism under a strict interpretation of Koch's postulates. The use of nonculture-based techniques such as molecular genetics has allowed for alternative establishment of causation.

While the causative organisms have to date been impossible to culture *in vitro*, it has been possible to grow them in animals such as mice and armadillos.

Naturally occurring infection also has been reported in nonhuman primates, including the African chimpanzee, sooty mangabey, and cynomolgus macaque, as well as in armadillos and red squirrels.

Risk Factors

The greatest risk factor for developing leprosy is contact with another case of leprosy. Contacts of people with leprosy are five to eight times more likely to develop leprosy than members of the general population. Other risk factors are poorly understood. However, conditions that reduce immune function, such as malnutrition, other illnesses, or host genetic differences, may increase the risk of developing leprosy. Despite this, infection with HIV does not appear to increase the risk of developing leprosy.

Transmission

Transmission of leprosy occurs during close contact with those who are infected. Transmission is believed to be by nasal droplets.

Leprosy is not known to be either sexually transmitted or highly infectious. People are no longer infectious after as little as two weeks of treatment.

Leprosy may also be transmitted to humans by armadillos and may be present in three species of nonhuman primates.

Two exit routes of *M. leprae* from the human body often described are the skin and the nasal mucosa, although their relative importance is not clear. Lepromatous cases show large numbers of organisms deep in the dermis, but whether they reach the skin surface in sufficient numbers is doubtful.

The skin and the upper respiratory tract are most likely entry route. While older research dealt with the skin route, recent research has increasingly favored the respiratory route. Experimental transmission of leprosy through aerosols containing *M. leprae* in immunosuppressed mice was accomplished, suggesting a similar possibility in humans.

Genetics

Name	Locus	OMIM	Gene
LPRS1	10p13	609888	
LPRS2	6q25	607572	*PARK2, PACRG*
LPRS3	4q32	246300	*TLR2*
LPRS4	6p21.3	610988	*LTA*
LPRS5	4p14	613223	*TLR1*
LPRS6	13q14.11	613407	

Several genes have been associated with a susceptibility to leprosy. Many people's immune systems are able to eliminate leprosy during the early infection stage before severe symptoms develop. A defect in cell-mediated immunity may cause susceptibility to leprosy. The region of DNA responsible for this variability is also involved in Parkinson's disease, giving rise to current speculation that the two disorders may be linked in some way at the biochemical level. Some evidence indicates not all people who are infected with *M. leprae* develop leprosy, and genetic factors have long been thought to play a role, due to the observation of clustering of leprosy around certain families, and the failure to understand why certain individuals develop lepromatous leprosy while others develop other types of leprosy.

Pathophysiology

How the infection produces the symptoms of the disease is not known.

Diagnosis

According to the World Health Organization, diagnosis in areas where people are frequently infected is based on one of these main signs:

- Skin lesion consistent with leprosy and with definite sensory loss

- Positive skin smears

Skin lesions can be single or multiple, and usually hypopigmented, although occasionally reddish or copper-colored. The lesions may be macules (flat), papules (raised), or nodular. Sensory loss at the skin lesion is important because this feature can help differentiate it from other causes of skin lesions such as tinea versicolor. Thickened nerves are associated with leprosy and can be accompanied by loss of sensation or muscle weakness. However, without the characteristic skin lesion and sensory loss, muscle weakness is not considered a reliable sign of leprosy.

Positive skin smears: In some case, acid-fast leprosy bacilli are considered diagnostic; however, the diagnosis is clinical.

Diagnosis in areas where the disease is uncommon, such as the United States, is often delayed because healthcare providers are unaware of leprosy and its symptoms. Early diagnosis and treatment prevent nerve involvement, the hallmark of leprosy, and the disability it causes.

Many kinds of leprosy are known, but some symptoms are common, including runny nose, dry scalp, eye problems, skin lesions, muscle weakness, reddish skin, smooth, shiny, diffuse thickening of facial skin, ear, and hand, loss of sensation in fingers and toes, thickening of peripheral nerves, and flat nose due to destruction of nasal cartilage. Also, phonation and resonation of sound occurs during speech. Often, atrophy of the testes and impotency happen.

Classification

Several different approaches for classifying leprosy exist, but parallels exist.

- The World Health Organization system distinguishes "paucibacillary" and "multibacillary" based upon the proliferation of bacteria.("pauci-" refers to a low quantity.)

- The SHAY scale provides five gradations.

- The ICD-10, though developed by the WHO, uses Ridley-Jopling and not the WHO system. It also adds an indeterminate ("I") entry.

- In MeSH, three groupings are used.

WHO	Ridley-Jopling	ICD-10	MeSH	Description	Lepromin test
Paucibacillary	tuberculoid ("TT"), borderline tuberculoid ("BT")	A30.1, A30.2	Tuberculoid	It is characterized by one or more hypopigmented skin macules and patches where skin sensations are lost because of damaged peripheral nerves that have been attacked by the human host's immune cells.	Positive
Multibacillary	midborderline or borderline ("BB")	A30.3	Borderline	Borderline leprosy is of intermediate severity and is the most common form. Skin lesions resemble tuberculoid leprosy, but are more numerous and irregular; large patches may affect a whole limb, and peripheral nerve involvement with weakness and loss of sensation is common. This type is unstable and may become more like lepromatous leprosy or may undergo a reversal reaction, becoming more like the tuberculoid form.	
Multibacillary	borderline lepromatous ("BL"), and lepromatous ("LL")	A30.4, A30.5	Lepromatous	It is associated with symmetric skin lesions, nodules, plaques, thickened dermis, and frequent involvement of the nasal mucosa resulting in nasal congestion and nose bleeds, but, typically, detectable nerve damage is late.	Negative

A difference in immune response to the tuberculoid and lepromatous forms is seen.

Leprosy may also be divided into:

- Early and indeterminate leprosy

- Tuberculoid leprosy

- Borderline tuberculoid leprosy

- Borderline leprosy

- Borderline lepromatous leprosy

- Lepromatous leprosy

- Histoid leprosy

- Diffuse leprosy of Lucio and Latapí

This disease may also occur with only neural involvement, without skin lesions.

Prevention

Early detection of the disease is important, since physical and neurological damage may be irreversible even if cured. Medications can decrease the risk of those living with people with leprosy from acquiring the disease and likely those with whom people with leprosy come into contact outside the home. However, concerns are known of resistance, cost, and disclosure of a person's infection status when doing follow-up of contacts. Therefore, the WHO recommends that people who live in the same household be examined for leprosy and only be treated if symptoms are present.

The Bacillus Calmette–Guérin (BCG) vaccine offers a variable amount of protection against leprosy in addition to tuberculosis. It appears to be 26 to 41% effective (based on controlled trials) and about 60% effective based on observational studies with two doses possibly working better than one. Development of a more effective vaccine is ongoing,

Treatment

MDT antileprosy drugs: standard regimens

A number of leprostatic agents are available for treatment. For paucibacillary (PB or tuberculoid) cases, treatment with daily dapsone and monthly rifampicin for six months is recommended. While for multibacillary (MB or lepromatous) cases, treatment with daily dapsone and clofazimine along with monthly rifampicin for 12 months is recommended.

Multidrug therapy (MDT) remains highly effective, and people are no longer infectious after the first monthly dose. It is safe and easy to use under field conditions due to its presentation in calendar blister packs. Relapse rates remain low, and no resistance to the combined drugs is seen.

Epidemiology

In 2012, the number of cases of leprosy was about 180,000. In 2011, the approximate number of new leprosy cases diagnosed was 220,000.

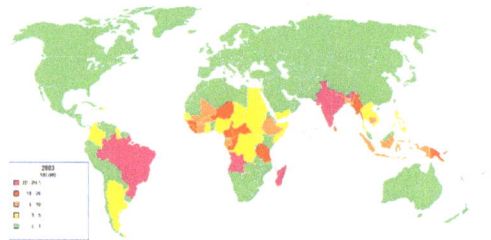

World distribution of leprosy, 2003

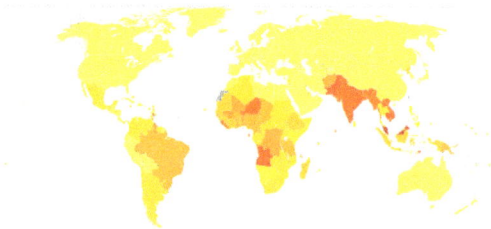

Disability-adjusted life year for leprosy per 100,000 inhabitants in 2004

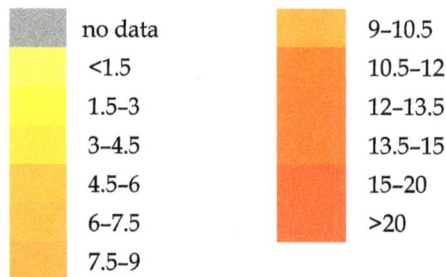

no data		9–10.5	
<1.5		10.5–12	
1.5–3		12–13.5	
3–4.5		13.5–15	
4.5–6		15–20	
6–7.5		>20	
7.5–9			

As of 2013, 14 countries contain 95% of the globally reported leprosy cases. Of these, India has the greatest number of cases (59%), followed by Brazil (14%) and Indonesia (8%). Although the number of cases worldwide continues to fall, pockets of high prevalence remain in certain areas such as Brazil, South Asia (India, Nepal, Bhutan), some parts of Africa (Tanzania, Madagascar, Mozambique), and the western Pacific.

The number of cases of leprosy was in the tens of millions in the 1960s, a series of national (the International Federation of Anti-Leprosy Associations) and international (the WHO's "Global Strategy for Reducing Disease Burden Due to Leprosy") initiatives have reduced the total number and the number of new cases of the disease. In 1995, two to three million people were estimated to be permanently disabled because of leprosy.

Disease Burden

Although the number of new leprosy cases occurring each year is important as a measure of transmission, it is difficult to measure due to leprosy's long incubation period, delays in diagnosis after onset of the disease, and the lack of laboratory tools to detect it in the very early stages. Instead, the registered prevalence is used. Registered prevalence is a useful proxy indicator of the disease burden, as it reflects the number of active leprosy cases diagnosed with the disease and receiving treatment with MDT at a given point in time. The prevalence rate is defined as the number of cases registered for MDT treatment among the population in which the cases have occurred, again at a given point in time.

New case detection is another indicator of the disease that is usually reported by countries on an annual basis. It includes cases diagnosed with onset of disease in the year in question (true incidence) and a large proportion of cases with onset in previous years (termed a backlog prevalence of undetected cases).

Endemic countries also report the number of new cases with established disabilities at the time of detection, as an indicator of the backlog prevalence. Determination of the time of onset of the disease is, in general, unreliable, is very labor-intensive, and is seldom done in recording these statistics.

History

G. H. A. Hansen, discoverer of *M. leprae*

Using comparative genomics, in 2005, geneticists traced the origins and worldwide distribution of leprosy from East Africa or the Near East along human migration routes. They found four strains of *M. leprae* with specific regional locations. Strain 1 occurs predominately in Asia, the Pacific region, and East Africa; strain 4, in West Africa and the Caribbean; strain 3 in Europe, North Africa, and the Americas; and strain 2 only in Ethiopia, Malawi, Nepal/north India, and New Caledonia.

On the basis of this, they offer a map of the dissemination of leprosy in the world. This confirms the spread of the disease along the migration, colonisation, and slave trade routes taken from East Africa to India, West Africa to the New World, and from Africa into Europe and vice versa.

The oldest skeletal evidence for the disease was found in the human remains from the archaeological sites of Balathal and Harappa, in India and Pakistan, respectively.

Although retrospectively identifing descriptions of leprosy-like symptoms is difficult, what appears to be leprosy was discussed by Hippocrates in 460 BC. In 1846, Francis Adams produced *The Seven Books of Paulus Aegineta* which included a commentary on all medical and surgical knowledge and descriptions and remedies to do with leprosy from the Romans, Greeks, and Arabs.

Interpretations of the presence of leprosy have been made on the basis of descriptions in ancient Indian (Atharva Verda and Kausika Sutra), Greek, and Middle Eastern documentary sources that describe skin afflictions.

Skeletal remains from the second millennium BC, discovered in 2009, represent the oldest documented evidence for leprosy. Located at Balathal, in Rajasthan, northwest India, the discoverers suggest that if the disease did migrate from Africa, to India, during the third millennium BC "at a time when there was substantial interaction among the Indus Civilization, Mesopotamia, and Egypt, there needs to be additional skeletal and molecular evidence of leprosy in India and Africa so as to confirm the African origin of the disease." A proven human case was verified by DNA taken from the shrouded remains of a man discovered in a tomb next to the Old City of Jerusalem dated by radiocarbon methods to 1–50 AD.

The causative agent of leprosy, *M. leprae*, was discovered by G. H. Armauer Hansen in Norway in 1873, making it the first bacterium to be identified as causing disease in humans. The first effective treatment (promin) became available in the 1940s. In the 1950s, dapsone was introduced. The search for further effective antileprosy drugs led to the use of clofazimine and rifampicin in the 1960s and 1970s. Later, Indian scientist Shantaram Yawalkar and his colleagues formulated a combined therapy using rifampicin and dapsone, intended to mitigate bacterial resistance. MDT combining all three drugs was first recommended by the WHO] in 1981. These three antileprosy drugs are still used in the standard MDT regimens.

Leprosy was once believed to be highly contagious and was treated with mercury—all of which applied to syphilis, which was first described in 1530. Many early cases thought to be leprosy could actually have been syphilis. Resistance has developed to initial treatment. Until the introduction of MDT in the early 1980s, the disease could not be diagnosed and treated successfully within the community.

Japan still has sanatoriums (although Japan's sanatoriums no longer have active leprosy cases, nor are survivors held in them by law).

The importance of the nasal mucosa in the transmission of *M leprae* was recognized as early as 1898 by Schäffer, in particular that of the ulcerated mucosa.

Society and Culture

Two lepers denied entrance to town, 14th century

India

India was one of the first countries to have acted against leprosy. India enacted the Leprosy Act of 1898 which institutionalized those affected and segregated them by gender to prevent reproduction. The Act was difficult to enforce, but was only repealed in 1983 after MDT therapy became widely available. In 1983, the National Leprosy Elimination Programme, previously the National Leprosy Control Programme, changed its methods from surveillance to treatment of people with leprosy. India still accounts for over half of the global disease burden.

Treatment Cost

Between 1995 and 1999, the WHO, with the aid of the Nippon Foundation, supplied all endemic countries with free MDT in blister packs, channelled through ministries of health. This free provision was extended in 2000 and again in 2005 with donations by the MDT manufacturer Novartis through the WHO. In the latest agreement signed between the company and the WHO in October 2010, the provision of free MDT by the WHO to all endemic countries has run until the end of 2015. At the national level, nongovernment organizations affiliated with the national programme will continue to be provided with an appropriate free supply of this WHO-supplied MDT by the government.

Historical Texts

Written accounts of leprosy date back thousands of years. Various skin diseases translated as leprosy appear in the ancient Indian text, the *Atharava Veda*, as early as 2000 BC. Another Indian text, the *Laws of Manu* (1500 BC), prohibited contact with those infected with the disease and made marriage to a person infected with leprosy punishable.

Many English translations of the Bible translate *tzaraath* as "leprosy," a confusion that derives from the use of the koine cognate (which can mean any disease causing scaly skin) in the Septuagint. Ancient sources such as the Talmud (Sifra 63) make clear that *tzaraath* refers to various types of lesions or stains associated with ritual impurity and occurring on cloth, leather, or houses, as well as skin. It may sometimes be a symptom of the disease described in this article, but has many other causes, as well. The New Testament describes instances of Jesus healing people with leprosy (Luke 5:10), although the precise relationship between this, *tzaraath*, and Hansen's disease is not established.

The biblical perception that people with leprosy were unclean may be connected to a passage from Leviticus 13: 44-46. Judeo-Christian belief held that leprosy was a moral disease, and early Christians believed that those affected by leprosy were being punished by God for sinful behavior. Moral associations have persisted throughout history. Pope Gregory the Great (540-604) and Isidor of Seville (560-636) considered people with the disease to be heretics.

Middle Ages

It is believed that a rise in leprosy in Europe occurred in the Middle Ages based on the increased number of hospitals created to treat leprosy patients in the 12th and 13th centuries. France alone had nearly 2,000 leprosariums during this period.

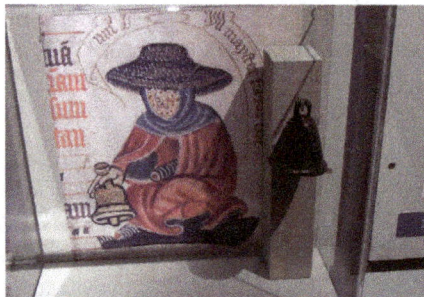

Medieval leper bell

The social perception in medieval communities was generally one of fear, and those people infected with the disease were thought to be unclean, untrustworthy, and morally corrupt. People with leprosy were also often required to wear clothing that identified them as such or carry a bell announcing their presence. Segregation from mainstream society was common. The third Lateran Council of 1179 and a 1346 edict by King

Edward expelled lepers from city limits. Because of the moral stigma of the disease, methods of treatment were both physical and spiritual, and leprosariums were established under the purview of the church.

Nineteenth Century

Norway

Norway was the location of a progressive stance on leprosy tracking and treatment, and played an influential role in European understanding of the disease. In 1832, Dr. JJ Hjort conducted the first leprosy survey, thus establishing a basis for epidemiological surveys. Subsequent surveys resulted in the establishment of a national leprosy registry to study the causes of leprosy and for tracking of rate of infection.

Early leprosy research throughout Europe was conducted by Norwegian scientists, Daniel Cornelius Danielssen and C.W. Boeck. Their work resulted in the establishment of the National Leprosy Research and Treatment Center. Danielssen and Boeck believed the cause of leprosy transmission was hereditary. This stance was influential in advocating for the isolation of those infected by gender to prevent reproduction.

Colonialism and Imperialism

Though leprosy in Europe was again on the decline by the 1860s, Western countries embraced isolation treatment out of fear of the spread of disease from developing countries, minimal understanding of bacteriology, lack of diagnostic ability or knowledge of how contagious the disease was, and missionary activity. Growing imperialism and pressures of the industrial revolution resulted in a Western presence in countries where leprosy was endemic, namely the British presence in India. Isolation treatment methods were observed by Surgeon-Mayor Henry Vandyke Carter of the British Colony in India while visiting Norway, and these methods were applied in India with the financial and logistical assistance of religious missionaries. Colonial and religious influence and associated stigma continued to be a major factor in the treatment and public perception of leprosy in endemic developing countries until the mid-20th century.

Stigma

Despite effective treatment and education efforts, leprosy stigma continues to be problematic in endemic developing countries. Leprosy is most prevalent amongst impoverished or marginalized populations where social stigma is likely to be compounded by other social inequities. Fears of ostracism, loss of employment, or expulsion from family and society may contribute to a delayed diagnosis and treatment.

Folk models of belief, lack of education, and religious connotations of the disease continue to influence social perceptions of those afflicted in many parts of the world. In Brazil, for example, folklore holds that leprosy is transmitted by dogs, it is a disease associated with sexual promiscuity, and is sometimes thought to be punishment for sins or moral transgressions. Socioeconomic factors also have a direct impact. Lower-class domestic workers who are often employed by those in a higher socioeconomic class may find their employment in jeopardy as physical manifestations of the disease become apparent. Skin discoloration and darker pigmentation resulting from the disease also has social repercussions.

In extreme cases in northern India, leprosy is equated with an "untouchable" status that "often persists long after (individuals with leprosy) have been cured of the disease, creating lifelong prospects of divorce, eviction, loss of employment, and ostracism from family and social networks."

Programs and Treatment

The WHO states that diagnosis and treatment with MDT is easy and effective, and a 45% decline in disease burden has occurred since MDT has become more widely available. The organization emphasizes the importance of fully integrating leprosy treatment into public health services, effective diagnosis and treatment, and access to information.

In some instances in India, community based rehabilitation is embraced by local governments and NGOs alike. Often, the identity cultivated by a community environment is preferable to reintegration, and models of self-management and collective agency independent of NGOs and government support have been desirable and successful.

Acute Prostatitis

Acute prostatitis is a serious bacterial infection of the prostate gland. This infection is a medical emergency. It should be distinguished from other forms of prostatitis such as chronic bacterial prostatitis and chronic pelvic pain syndrome (CPPS).

Signs and Symptoms

Men with this disease often have chills, fever, pain in the lower back and genital area, urinary frequency and urgency often at night, burning or painful urination, body aches, and a demonstrable infection of the urinary tract, as evidenced by white blood cells and bacteria in the urine. Acute prostatitis may be a complication of prostate biopsy. Often, the prostate gland is very tender to palpation through the rectum.

Diagnosis

Acute prostatitis is relatively easy to diagnose due to its symptoms that suggest infection. The organism may be found in blood or urine, and sometimes in both. Common bacteria are *Escherichia coli, Klebsiella, Proteus, Pseudomonas, Enterobacter,*

Enterococcus, Serratia, and *Staphylococcus aureus*. This can be a medical emergency in some patients and hospitalization with intravenous antibiotics may be required. A complete blood count reveals increased white blood cells. Sepsis from prostatitis is very rare, but may occur in immunocompromised patients; high fever and malaise generally prompt blood cultures, which are often positive in sepsis. A prostate massage should never be done in a patient with suspected acute prostatitis, since it may induce sepsis. Since bacteria causing the prostatitis is easily recoverable from the urine, prostate massage is not required to make the diagnosis. Rectal palpation usually reveals an enlarged, exquisitely tender, swollen prostate gland, which is firm, warm, and, occasionally, irregular to the touch. C-reactive protein is elevated in most cases.

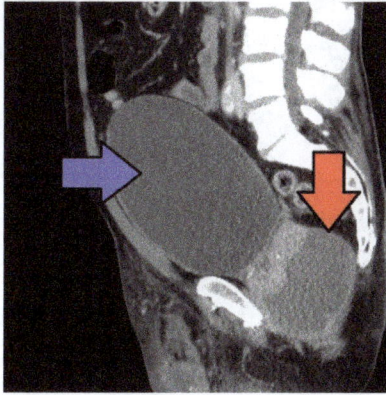

Abscess of the prostate resulting in urinary retention

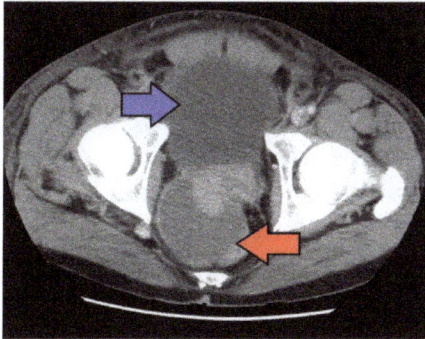

Abscess of the prostate resulting in urinary retention

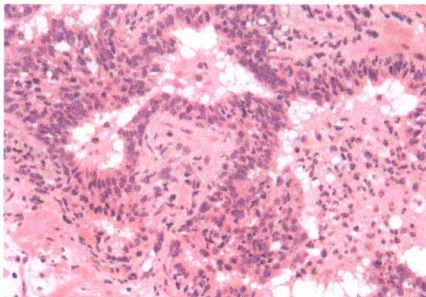

Micrograph showing a neutrophilic infiltration of prostatic glands - the histologic correlate of acute prostatitis. H&E stain.

Prostate biopsies are not indicated as the (clinical) features (described above) are diagnostic. The histologic correlate of acute prostatitis is a neutrophilic infiltration of the prostate gland.

Acute prostatitis is associated with a transiently elevated PSA, i.e., the PSA is increased during an episode of acute prostatitis and then decreases again after it has resolved. PSA testing is not indicated in the context of uncomplicated acute prostatitis.

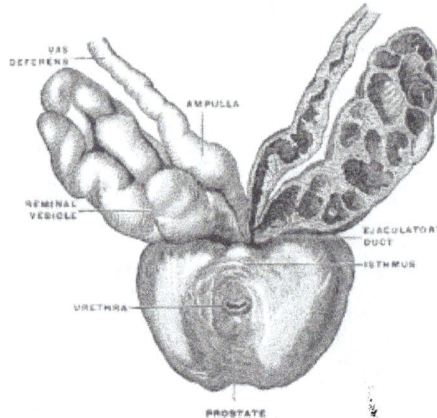

Prostate, urethra, and seminal vesicles.

Treatment

Antibiotics are the first line of treatment in acute prostatitis (Cat. I). Antibiotics usually resolve acute prostatitis infections in a very short time, however a minimum of two to four weeks of therapy is recommended to eradicate the offending organism completely. Appropriate antibiotics should be used, based on the microbe causing the infection. Some antibiotics have very poor penetration of the prostatic capsule, others, such as Ciprofloxacin, Co-trimoxazole and tetracyclines such as doxycycline penetrate well. In acute prostatitis, penetration of the prostate is not as important as for category II because the intense inflammation disrupts the prostate-blood barrier. It is more important to choose a bacteriocidal antibiotic (kills bacteria, e.g. quinolone) rather than a bacteriostatic antibiotic (slows bacterial growth, e.g. tetracycline) for acute potentially life-threatening infections. Severely ill patients may need hospitalization, while nontoxic patients can be treated at home with bed rest, analgesics, stool softeners, and hydration. Patients in urinary retention are best managed with a suprapubic catheter or intermittent catheterization. Lack of clinical response to antibiotics should raise the suspicion of an abscess and prompt an imaging study such as a transrectal ultrasound (TRUS).

Prognosis

Full recovery without sequelae is usual.

Tuberculosis

Tuberculosis (TB) is an infectious disease caused by the bacterium *Mycobacterium tuberculosis* (MTB). Tuberculosis generally affects the lungs, but can also affect other parts of the body. Most infections do not have symptoms, known as latent tuberculosis. About 10% of latent infections progress to active disease which, if left untreated, kills about half of those infected. The classic symptoms of active TB are a chronic cough with blood-containing sputum, fever, night sweats, and weight loss. The historical term "consumption" came about due to the weight loss. Infection of other organs can cause a wide range of symptoms.

Tuberculosis is spread through the air when people who have active TB in their lungs cough, spit, speak, or sneeze. People with latent TB do not spread the disease. Active infection occurs more often in people with HIV/AIDS and in those who smoke. Diagnosis of active TB is based on chest X-rays, as well as microscopic examination and culture of body fluids. Diagnosis of latent TB relies on the tuberculin skin test (TST) or blood tests.

Prevention of TB involves screening those at high risk, early detection and treatment of cases, and vaccination with the bacillus Calmette-Guérin vaccine. Those at high risk include household, workplace, and social contacts of people with active TB. Treatment requires the use of multiple antibiotics over a long period of time. Antibiotic resistance is a growing problem with increasing rates of multiple drug-resistant tuberculosis (MDR-TB).

One-third of the world's population is thought to be infected with TB. New infections occur in about 1% of the population each year. In 2014, there were 9.6 million cases of active TB which resulted in 1.5 million deaths. More than 95% of deaths occurred in developing countries. The number of new cases each year has decreased since 2000. About 80% of people in many Asian and African countries test positive while 5–10% of people in the United States population tests positive by the tuberculin test. Tuberculosis has been present in humans since ancient times.

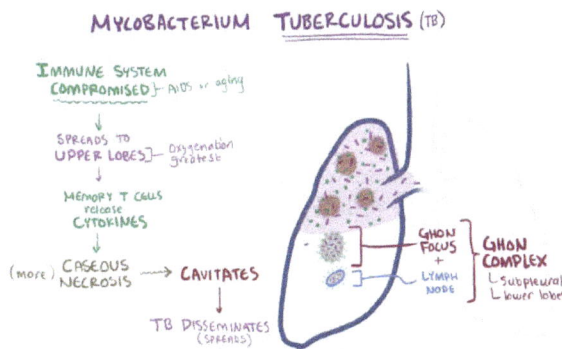

Video explanation

Signs and Symptoms

Tuberculosis may infect any part of the body, but most commonly occurs in the lungs (known as pulmonary tuberculosis). Extrapulmonary TB occurs when tuberculosis develops outside of the lungs, although extrapulmonary TB may coexist with pulmonary TB.

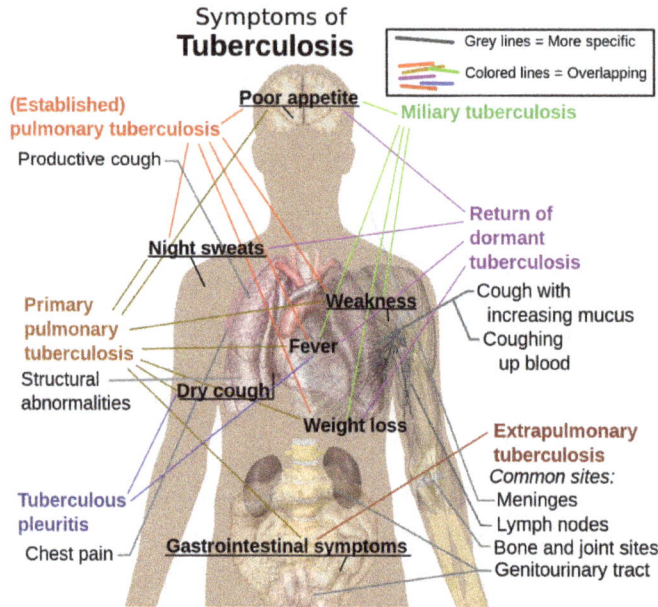

The main symptoms of variants and stages of tuberculosis are given, with many symptoms overlapping with other variants, while others are more (but not entirely) specific for certain variants. Multiple variants may be present simultaneously.

General signs and symptoms include fever, chills, night sweats, loss of appetite, weight loss, and fatigue. Significant nail clubbing may also occur.

Pulmonary

If a tuberculosis infection does become active, it most commonly involves the lungs (in about 90% of cases). Symptoms may include chest pain and a prolonged cough producing sputum. About 25% of people may not have any symptoms (i.e. they remain "asymptomatic"). Occasionally, people may cough up blood in small amounts, and in very rare cases, the infection may erode into the pulmonary artery or a Rasmussen's aneurysm, resulting in massive bleeding. Tuberculosis may become a chronic illness and cause extensive scarring in the upper lobes of the lungs. The upper lung lobes are more frequently affected by tuberculosis than the lower ones. The reason for this difference is not clear. It may be due to either better air flow, or poor lymph drainage within the upper lungs.

Extrapulmonary

In 15–20% of active cases, the infection spreads outside the lungs, causing other kinds of TB. These are collectively denoted as "extrapulmonary tuberculosis".

Extrapulmonary TB occurs more commonly in immunosuppressed persons and young children. In those with HIV, this occurs in more than 50% of cases. Notable extrapulmonary infection sites include the pleura (in tuberculous pleurisy), the central nervous system (in tuberculous meningitis), the lymphatic system (in scrofula of the neck), the genitourinary system (in urogenital tuberculosis), and the bones and joints (in Pott disease of the spine), among others. When it spreads to the bones, it is also known as "osseous tuberculosis", a form of osteomyelitis. Sometimes, bursting of a tubercular abscess through skin results in tuberculous ulcer. An ulcer originating from nearby infected lymph nodes is painless, slowly enlarging and has an appearance of "wash leather". A potentially more serious, widespread form of TB is called "disseminated tuberculosis", also known as miliary tuberculosis. Miliary TB makes up about 10% of extrapulmonary cases.

Causes

Mycobacteria

The main cause of TB is *Mycobacterium tuberculosis*, a small, aerobic, nonmotile bacillus. The high lipid content of this pathogen accounts for many of its unique clinical characteristics. It divides every 16 to 20 hours, which is an extremely slow rate compared with other bacteria, which usually divide in less than an hour. Mycobacteria have an outer membrane lipid bilayer. If a Gram stain is performed, MTB either stains very weakly "Gram-positive" or does not retain dye as a result of the high lipid and mycolic acid content of its cell wall. MTB can withstand weak disinfectants and survive in a dry state for weeks. In nature, the bacterium can grow only within the cells of a host organism, but *M. tuberculosis* can be cultured in the laboratory.

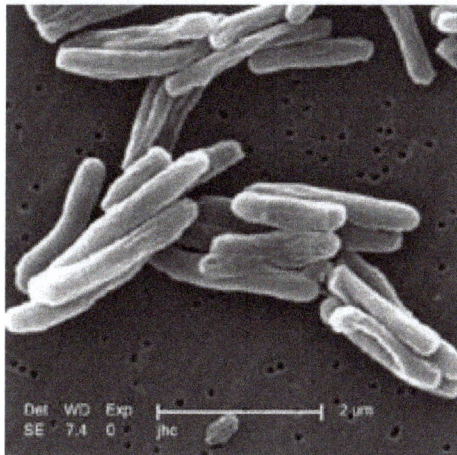

Scanning electron micrograph of *M. tuberculosis*

Using histological stains on expectorated samples from phlegm (also called "sputum"), scientists can identify MTB under a microscope. Since MTB retains certain stains even after being treated with acidic solution, it is classified as an acid-fast bacillus. The most common acid-fast staining techniques are the Ziehl–Neelsen stain and the Kinyoun stain, which dye acid-fast bacilli a bright red that stands out against a blue background. Auramine-rhodamine staining and fluorescence microscopy are also used.

The *M. tuberculosis* complex (MTBC) includes four other TB-causing mycobacteria: *M. bovis*, *M. africanum*, *M. canetti*, and *M. microti*. *M. africanum* is not widespread, but it is a significant cause of tuberculosis in parts of Africa. *M. bovis* was once a common cause of tuberculosis, but the introduction of pasteurized milk has almost completely eliminated this as a public health problem in developed countries. *M. canetti* is rare and seems to be limited to the Horn of Africa, although a few cases have been seen in African emigrants. *M. microti* is also rare and is seen almost only in immunodeficient people, although its prevalence may be significantly underestimated.

Other known pathogenic mycobacteria include *M. leprae*, *M. avium*, and *M. kansasii*. The latter two species are classified as "nontuberculous mycobacteria" (NTM). NTM cause neither TB nor leprosy, but they do cause pulmonary diseases that resemble TB.

Risk Factors

A number of factors make people more susceptible to TB infections. The most important risk factor globally is HIV; 13% of all people with TB are infected by the virus. This is a particular problem in sub-Saharan Africa, where rates of HIV are high. Of people without HIV who are infected with tuberculosis, about 5–10% develop active disease during their lifetimes; in contrast, 30% of those coinfected with HIV develop the active disease.

Tuberculosis is closely linked to both overcrowding and malnutrition, making it one of the principal diseases of poverty. Those at high risk thus include: people who inject illicit drugs, inhabitants and employees of locales where vulnerable people gather (e.g. prisons and homeless shelters), medically underprivileged and resource-poor communities, high-risk ethnic minorities, children in close contact with high-risk category patients, and health-care providers serving these patients.

Chronic lung disease is another significant risk factor. Silicosis increases the risk about 30-fold. Those who smoke cigarettes have nearly twice the risk of TB compared to non-smokers.

Other disease states can also increase the risk of developing tuberculosis. These include alcoholism and diabetes mellitus (three-fold increase).

Certain medications, such as corticosteroids and infliximab (an anti-αTNF monoclonal antibody), are becoming increasingly important risk factors, especially in the developed world.

Genetic susceptibility also exists, for which the overall importance remains undefined.

Mechanism

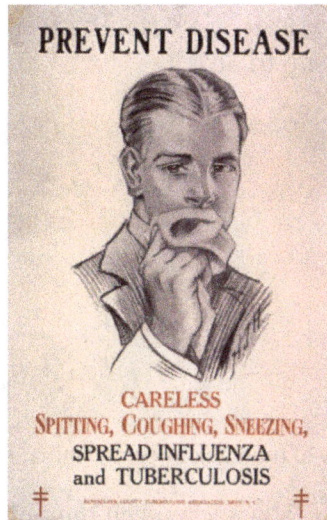

Public health campaigns in the 1920s tried to halt the spread of TB.

Transmission

When people with active pulmonary TB cough, sneeze, speak, sing, or spit, they expel infectious aerosol droplets 0.5 to 5.0 μm in diameter. A single sneeze can release up to 40,000 droplets. Each one of these droplets may transmit the disease, since the infectious dose of tuberculosis is very small (the inhalation of fewer than 10 bacteria may cause an infection).

People with prolonged, frequent, or close contact with people with TB are at particularly high risk of becoming infected, with an estimated 22% infection rate. A person with active but untreated tuberculosis may infect 10–15 (or more) other people per year. Transmission should occur from only people with active TB – those with latent infection are not thought to be contagious. The probability of transmission from one person to another depends upon several factors, including the number of infectious droplets expelled by the carrier, the effectiveness of ventilation, the duration of exposure, the virulence of the *M. tuberculosis* strain, the level of immunity in the uninfected person, and others. The cascade of person-to-person spread can be circumvented by segregating those with active ("overt") TB and putting them on anti-TB drug regimens. After about two weeks of effective treatment, subjects with nonresistant active infections generally do not remain contagious to others. If someone does become infected, it typically takes three to four weeks before the newly infected person becomes infectious enough to transmit the disease to others.

Pathogenesis

About 90% of those infected with *M. tuberculosis* have asymptomatic, latent TB infections (sometimes called LTBI), with only a 10% lifetime chance that the latent infection will progress to overt, active tuberculous disease. In those with HIV, the risk of developing active TB increases to nearly 10% a year. If effective treatment is not given, the death rate for active TB cases is up to 66%.

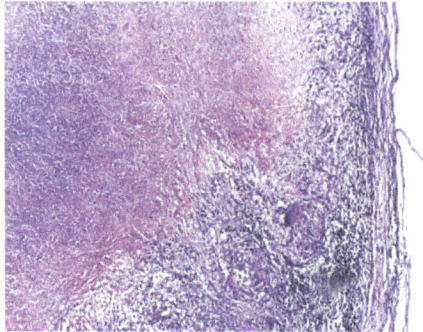

Microscopy of tuberculous epididymitis. H&E stain

TB infection begins when the mycobacteria reach the pulmonary alveoli, where they invade and replicate within endosomes of alveolar macrophages. Macrophages identify the bacterium as foreign and attempt to eliminate it by phagocytosis. During this process, the bacterium is enveloped by the macrophage and stored temporarily in a membrane-bound vesicle called a phagosome. The phagosome then combines with a lysosome to create a phagolysosome. In the phagolysosome, the cell attempts to use reactive oxygen species and acid to kill the bacterium. However, *M. tuberculosis* has a thick, waxy mycolic acid capsule that protects it from these toxic substances. *M. tuberculosis* is able to reproduce inside the macrophage and will eventually kill the immune cell.

The primary site of infection in the lungs, known as the "Ghon focus", is generally located in either the upper part of the lower lobe, or the lower part of the upper lobe. Tuberculosis of the lungs may also occur via infection from the blood stream. This is known as a Simon focus and is typically found in the top of the lung. This hematogenous transmission can also spread infection to more distant sites, such as peripheral lymph nodes, the kidneys, the brain, and the bones. All parts of the body can be affected by the disease, though for unknown reasons it rarely affects the heart, skeletal muscles, pancreas, or thyroid.

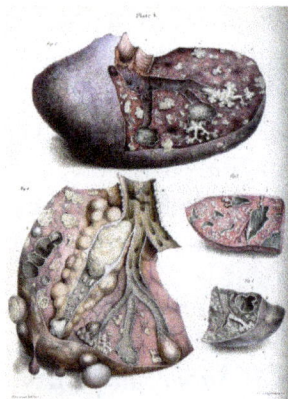

Robert Carswell's illustration of tubercle

Tuberculosis is classified as one of the granulomatous inflammatory diseases. Macrophages, T lymphocytes, B lymphocytes, and fibroblasts aggregate to form granulomas, with lymphocytes surrounding the infected macrophages. When other macrophages attack the infected macrophage, they fuse together to form a giant multinucleated cell in the alveolar lumen. The granuloma may prevent dissemination of the mycobacteria and provide a local environment for interaction of cells of the immune system. However, more recent evidence suggests that the bacteria use the granulomas to avoid destruction by the host's immune system. Macrophages and dendritic cells in the granulomas are unable to present antigen to lymphocytes; thus the immune response is suppressed. Bacteria inside the granuloma can become dormant, resulting in latent infection. Another feature of the granulomas is the development of abnormal cell death (necrosis) in the center of tubercles. To the naked eye, this has the texture of soft, white cheese and is termed caseous necrosis.

If TB bacteria gain entry to the blood stream from an area of damaged tissue, they can spread throughout the body and set up many foci of infection, all appearing as tiny, white tubercles in the tissues. This severe form of TB disease, most common in young children and those with HIV, is called miliary tuberculosis. People with this disseminated TB have a high fatality rate even with treatment (about 30%).

In many people, the infection waxes and wanes. Tissue destruction and necrosis are often balanced by healing and fibrosis. Affected tissue is replaced by scarring and cavities filled with caseous necrotic material. During active disease, some of these cavities are joined to the air passages bronchi and this material can be coughed up. It contains living bacteria, so can spread the infection. Treatment with appropriate antibiotics kills bacteria and allows healing to take place. Upon cure, affected areas are eventually replaced by scar tissue.

Diagnosis

M. tuberculosis (stained red) in sputum

Active Tuberculosis

Diagnosing active tuberculosis based only on signs and symptoms is difficult, as is diagnosing the disease in those who are immunosuppressed. A diagnosis of TB should, however, be considered in those with signs of lung disease or constitutional symptoms lasting longer than two weeks. A chest X-ray and multiple sputum cultures for acid-fast bacilli are typically part of the initial evaluation. Interferon-γ release assays and tuberculin skin tests are of little use in the developing world. IGRA have similar limitations in those with HIV.

A definitive diagnosis of TB is made by identifying *M. tuberculosis* in a clinical sample (e.g., sputum, pus, or a tissue biopsy). However, the difficult culture process for this slow-growing organism can take two to six weeks for blood or sputum culture. Thus, treatment is often begun before cultures are confirmed.

Nucleic acid amplification tests and adenosine deaminase testing may allow rapid diagnosis of TB. These tests, however, are not routinely recommended, as they rarely alter how a person is treated. Blood tests to detect antibodies are not specific or sensitive, so they are not recommended.

Latent Tuberculosis

The Mantoux tuberculin skin test is often used to screen people at high risk for TB. Those who have been previously immunized may have a false-positive test result. The test may be falsely negative in those with sarcoidosis, Hodgkin's lymphoma, malnutrition, and most notably, active tuberculosis. Interferon gamma release assays (IGRAs), on a blood sample, are recommended in those who are positive to the Mantoux test. These are not affected by immunization or most environmental mycobacteria, so they generate fewer false-positive results. However, they are affected by *M. szulgai, M. marinum,* and *M. kansasii.* IGRAs may increase sensitivity when used in addition to the skin test, but may be less sensitive than the skin test when used alone.

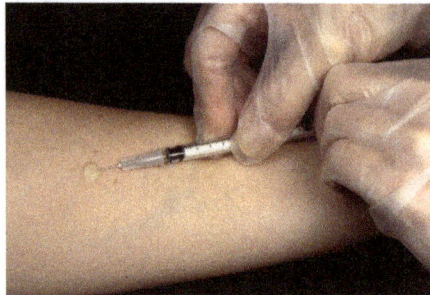

Mantoux tuberculin skin test

Prevention

Tuberculosis prevention and control efforts rely primarily on the vaccination of infants and the detection and appropriate treatment of active cases. The World Health Organization has achieved some success with improved treatment regimens, and a small decrease in case numbers. The US Preventive Services Task Force (USPSTF) recommends screening people who are at high risk for latent tuberculosis with either tuberculin skin tests or interferon-gamma release assays.

Vaccines

The only available vaccine as of 2011 is Bacillus Calmette-Guérin (BCG). In children it decreases the risk of getting the infection by 20% and the risk of infection turning into disease by nearly 60%.

It is the most widely used vaccine worldwide, with more than 90% of all children being vaccinated. The immunity it induces decreases after about ten years. As tuberculosis is uncommon in most of Canada, the United Kingdom, and the United States, BCG is administered to only those people at high risk. Part of the reasoning against the use of the vaccine is that it makes the tuberculin skin test falsely positive, reducing the test's use in screening. A number of new vaccines are currently in development.

Public Health

The World Health Organization declared TB a "global health emergency" in 1993, and in 2006, the Stop TB Partnership developed a Global Plan to Stop Tuberculosis that aimed to save 14 million lives between its launch and 2015. A number of targets they set were not achieved by 2015, mostly due to the increase in HIV-associated tuberculosis and the emergence of multiple drug-resistant tuberculosis. A tuberculosis classification system developed by the American Thoracic Society is used primarily in public health programs.

Management

Treatment of TB uses antibiotics to kill the bacteria. Effective TB treatment is difficult, due to the unusual structure and chemical composition of the mycobacterial cell wall, which hinders the entry of drugs and makes many antibiotics ineffective. The two antibiotics most commonly used are isoniazid and rifampicin, and treatments can be prolonged, taking several months. Latent TB treatment usually employs a single antibiotic, while active TB disease is best treated with combinations of several antibiotics to reduce the risk of the bacteria developing antibiotic resistance. People with latent infections are also treated to prevent them from progressing to active TB disease later in life. Directly observed therapy, i.e., having a health care provider watch the person take their medications, is recommended by the WHO in an effort to reduce the number of people not appropriately taking antibiotics. The evidence to support this practice over people simply taking their medications independently is poor. Methods to remind people of the importance of treatment do, however, appear effective.

New Onset

The recommended treatment of new-onset pulmonary tuberculosis, as of 2010, is six months of a combination of antibiotics containing rifampicin, isoniazid, pyrazinamide, and ethambutol for the first two months, and only rifampicin and isoniazid for the last four months. Where resistance to isoniazid is high, ethambutol may be added for the last four months as an alternative.

Recurrent Disease

If tuberculosis recurs, testing to determine to which antibiotics it is sensitive is important before determining treatment. If multiple drug-resistant TB is detected, treatment with at least four effective antibiotics for 18 to 24 months is recommended.

Medication Resistance

Primary resistance occurs when a person becomes infected with a resistant strain of TB. A person with fully susceptible MTB may develop secondary (acquired) resistance during therapy because of inadequate treatment, not taking the prescribed regimen appropriately (lack of compliance), or using low-quality medication. Drug-resistant TB is a serious public health issue in many developing countries, as its treatment is longer and requires more expensive drugs. MDR-TB is defined as resistance to the two most effective first-line TB drugs: rifampicin and isoniazid. Extensively drug-resistant TB is also resistant to three or more of the six classes of second-line drugs. Totally drug-resistant TB is resistant to all currently used drugs. It was first observed in 2003 in Italy, but not widely reported until 2012, and has also been found in Iran and India. Bedaquiline is tentatively supported for use in multiple drug-resistant TB.

XDR-TB is a term sometimes used to define *extensively resistant* TB, and constitutes one in ten cases of MDR-TB. Cases of XDR TB have been identified in more than 90% of countries.

Prognosis

Progression from TB infection to overt TB disease occurs when the bacilli overcome the immune system defenses and begin to multiply. In primary TB disease (some 1–5% of cases), this occurs soon after the initial infection. However, in the majority of cases, a latent infection occurs with no obvious symptoms. These dormant bacilli produce active tuberculosis in 5–10% of these latent cases, often many years after infection.

The risk of reactivation increases with immunosuppression, such as that caused by infection with HIV. In people coinfected with *M. tuberculosis* and HIV, the risk of reactivation increases to 10% per year. Studies using DNA fingerprinting of *M. tuberculosis* strains have shown reinfection contributes more substantially to recurrent TB than previously thought, with estimates that it might account for more than 50% of reactivated cases in areas where TB is common. The chance of death from a case of tuberculosis is about 4% as of 2008, down from 8% in 1995.

Epidemiology

Roughly one-third of the world's population has been infected with *M. tuberculosis*, with new infections occurring in about 1% of the population each year. However, most infections with *M. tuberculosis* do not cause TB disease, and 90–95% of infections remain asymptomatic. In 2012, an estimated 8.6 million chronic cases were active. In 2010, 8.8 million new cases of TB were diagnosed, and 1.20–1.45 million deaths occurred, most of these occurring in developing countries. Of these 1.45 million deaths, about 0.35 million occur in those also infected with HIV.

Tuberculosis is the second-most common cause of death from infectious disease (after those due to HIV/AIDS). The total number of tuberculosis cases has been decreasing since 2005, while new cases have decreased since 2002. China has achieved particularly dramatic progress, with about an 80% reduction in its TB mortality rate between 1990 and 2010. The number of new cases has declined by 17% between 2004–2014. Tuberculosis is more common in developing countries; about 80% of the population in many Asian and African countries test positive in tuberculin tests, while only 5–10% of the US population test positive. Hopes of totally controlling the disease have been dramatically dampened because of a number of factors, including the difficulty of developing an effective vaccine, the expensive and time-consuming diagnostic process, the necessity of many months of treatment, the increase in HIV-associated tuberculosis, and the emergence of drug-resistant cases in the 1980s.

In 2007, the country with the highest estimated incidence rate of TB was Swaziland, with 1,200 cases per 100,000 people. India had the largest total incidence, with an estimated 2.0 million new cases. In developed countries, tuberculosis is less common and is found mainly in urban areas. Rates per 100,000 people in different areas of the world were: globally 178, Africa 332, the Americas 36, Eastern Mediterranean 173, Europe 63, Southeast Asia 278, and Western Pacific 139 in 2010. In Canada and Australia, tuberculosis is many times more common among the aboriginal peoples, especially in remote areas. In the United States Native Americans have a fivefold greater mortality from TB, and racial and ethnic minorities accounted for 84% of all reported TB cases.

The rates of TB varies with age. In Africa, it primarily affects adolescents and young adults. However, in countries where incidence rates have declined dramatically (such as the United States), TB is mainly a disease of older people and the immunocompromised (risk factors are listed above). Worldwide, 22 "high-burden" states or countries together experience 80% of cases as well as 83% of deaths.

History

Tuberculosis has been present in humans since antiquity. The earliest unambiguous detection of *M. tuberculosis* involves evidence of the disease in the remains of bison in Wyoming dated to around 17,000 years ago. However, whether tuberculosis originated in bovines, then was transferred to humans, or whether it diverged from a common ancestor, is currently unclear. A comparison of the genes of *M. tuberculosis* complex (MTBC) in humans to MTBC in animals suggests humans did not acquire MTBC from animals during animal domestication, as was previously believed. Both strains of the tuberculosis bacteria share a common ancestor, which could have infected humans even before the Neolithic Revolution.

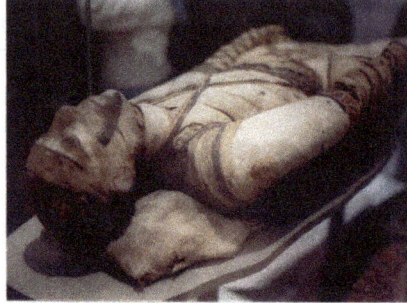

Egyptian mummy in the British Museum – tubercular decay has been found in the spine

Skeletal remains show prehistoric humans (4000 BC) had TB, and researchers have found tubercular decay in the spines of Egyptian mummies dating from 3000–2400 BC. Genetic studies suggest TB was present in the Americas from about 100 AD.

Before the Industrial Revolution, folklore often associated tuberculosis with vampires. When one member of a family died from it, the other infected members would lose their health slowly. People believed this was caused by the original person with TB draining the life from the other family members.

Although the pulmonary form associated with tubercles was established as a pathology by Dr Richard Morton in 1689, due to the variety of its symptoms, TB was not identified as a single disease until the 1820s. It was not named "tuberculosis" until 1839, by J. L. Schönlein.

During 1838–1845, Dr. John Croghan, the owner of Mammoth Cave, brought a number of people with tuberculosis into the cave in the hope of curing the disease with the constant temperature and purity of the cave air; they died within a year. Hermann Brehmer opened the first TB sanatorium in 1859 in Görbersdorf (now Sokołowsko), Silesia.

Dr. Robert Koch discovered the tuberculosis bacillus.

The bacillus causing tuberculosis, *M. tuberculosis*, was identified and described on 24 March 1882 by Robert Koch. He received the Nobel Prize in physiology or medicine in 1905 for this discovery. Koch did not believe the bovine (cattle) and human tuberculosis diseases were similar, which delayed the recognition of infected milk as a source of infection. Later, the risk of transmission from this source was dramatically reduced by the invention of the pasteurization process. Koch announced a glycerine extract of the tubercle bacilli as a "remedy" for tuberculosis in 1890, calling it "tuberculin". While it was not effective, it was later successfully adapted as a screening test for the presence of pre-symptomatic tuberculosis. The World Tuberculosis Day was established on 24 March for this reason.

Albert Calmette and Camille Guérin achieved the first genuine success in immunization against tuberculosis in 1906, using attenuated bovine-strain tuberculosis. It was called bacille Calmette–Guérin (BCG). The BCG vaccine was first used on humans in 1921 in France, but received widespread acceptance in the US, Great Britain, and Germany only after World War II.

Tuberculosis caused the most widespread public concern in the 19th and early 20th centuries as an endemic disease of the urban poor. In 1815, one in four deaths in England was due to "consumption". By 1918, one in six deaths in France was still caused by TB. After TB was determined to be contagious, in the 1880s, it was put on a notifiable disease list in Britain; campaigns were started to stop people from spitting in public places, and the infected poor were "encouraged" to enter sanatoria that resembled prisons (the sanatoria for the middle and upper classes offered excellent care and constant medical attention). Whatever the (purported) benefits of the "fresh air" and labor in the sanatoria, even under the best conditions, 50% of those who entered died within five years (*circa* 1916).

In Europe, rates of tuberculosis began to rise in the early 1600s to a peak level in the 1800s, when it caused nearly 25% of all deaths. By the 1950s, mortality had decreased nearly 90%. Improvements in public health began significantly reducing rates of tuberculosis even before the arrival of streptomycin and other antibiotics, although the disease remained a significant threat to public health such that when the Medical Research Council was formed in Britain in 1913, its initial focus was tuberculosis research.

In 1946, the development of the antibiotic streptomycin made effective treatment and cure of TB a reality. Prior to the introduction of this drug, the only treatment (except sanatoria) was surgical intervention, including the "pneumothorax technique", which involved collapsing an infected lung to "rest" it and allow tuberculous lesions to heal.

Because of the emergence of MDR-TB, surgery has been re-introduced as an option within the generally accepted standard of care in treating TB infections. Current surgical interventions involve removal of pathological chest cavities ("bullae") in the lungs to reduce the number of bacteria and to increase the exposure of the remaining bacteria to drugs in the bloodstream, thereby simultaneously reducing the total bacterial load and increasing the effectiveness of systemic antibiotic therapy.

Hopes of completely eliminating TB (*cf.* smallpox) from the population were dashed after the rise of drug-resistant strains in the 1980s. The subsequent resurgence of tuberculosis resulted in the declaration of a global health emergency by the World Health Organization in 1993.

Society and Culture

Names

Phthisis (Φθισις) is a Greek word for consumption, an old term for pulmonary tuberculosis; around 460 BC, Hippocrates described phthisis as a disease of dry seasons. The abbreviation "TB" is short for *tubercle bacillus*.

"Consumption" was the most common nineteenth century English word for the disease. The Latin root "con" meaning "completely" is linked to "sumere" meaning "to take up from under." In The Life and Death of Mr. Badman by John Bunyan, the author calls consumption "the captain of all these men of death."

Public Health Efforts

The World Health Organization, Bill and Melinda Gates Foundation, and US government are subsidizing a fast-acting diagnostic tuberculosis test for use in low- and middle-income countries. In addition to being fast-acting, the test can determine if there is resistance to the antibiotic rifampicin which may indicate multi-drug resistant tuberculosis and is accurate in those who are also infected with HIV. Many resource-poor places as of 2011 have access to only sputum microscopy.

India had the highest total number of TB cases worldwide in 2010, in part due to poor disease management within the private and public health care sector. Programs such as the Revised National Tuberculosis Control Program are working to reduce TB levels amongst people receiving public health care.

A 2014 the EIU-healthcare report that the need to address apathy and urging for increased funding. The report cites among others Lucica Ditui "[TB] is like an orphan. It has been neglected even in countries with a high burden and often forgotten by donors and those investing in health interventions."

Slow progress has led to frustration, expressed by the executive director of the Global Fund to Fight AIDS, Tuberculosis and Malaria – Mark Dybul: "we have the tools to end TB as a pandemic and public health threat on the planet, but we are not doing it." Several international organizations are pushing for more transparency in treatment, and more countries are implementing mandatory reporting of cases to the government, although adherence is often sketchy. Commercial treatment providers may at times overprescribe second-line drugs as well as supplementary treatment, promoting demands for further regulations. The government of Brazil provides universal TB-care, which reduces this problem. Conversely, falling rates of TB-infection may not relate to the number of programs directed at reducing infection rates, but may be tied to increased level of education, income, and health of the population. Costs of the disease, as calculated by the World Bank in 2009 may exceed 150 billion USD per year in "high burden" countries. Lack of progress eradicating the disease may also be due to lack of patient follow-up – as among the 250M rural migrants in China.

Stigma

Slow progress in preventing the disease may in part be due to stigma associated with TB. Stigma may be due to the fear of transmission from affected individuals. This stigma may additionally arise due to links between TB and poverty, and in Africa, AIDS. Such stigmatization may be both real and perceived, for example; in Ghana individuals with TB are banned from attending public gatherings.

Stigma towards TB may result in delays in seeking treatment, lower treatment compliance, and family members keeping cause of death secret – allowing the disease to spread further. At odds is Russia, where stigma was associated with increased treatment compliance. TB stigma also affects socially marginalized individuals to a greater degree and varies between regions.

One way to decrease stigma may be through the promotion of "TB clubs", where those infected may share experiences and offer support, or through counseling. Some studies have shown TB education programs to be effective in decreasing stigma, and may thus be effective in increasing treatment adherence. Despite this, studies on the relationship between reduced stigma and mortality are lacking as of 2010, and similar efforts to decrease stigma surrounding AIDS have been minimally effective. Some have claimed the stigma to be worse than the disease, and healthcare providers may unintentionally reinforce stigma, as those with TB are often perceived as difficult or otherwise undesirable. A greater understanding of the social and cultural dimensions of tuberculosis may also help with stigma reduction.

Research

The BCG vaccine has limitations, and research to develop new TB vaccines is ongoing. A number of potential candidates are currently in phase I and II clinical trials. Two main approaches are being used to attempt to improve the efficacy of available vaccines. One approach involves adding a subunit vaccine to BCG, while the other strategy is attempting to create new and better live vaccines. MVA85A, an example of a subunit vaccine, currently in trials in South Africa, is based on a genetically modified vaccinia virus. Vaccines are hoped to play a significant role in treatment of both latent and active disease.

To encourage further discovery, researchers and policymakers are promoting new economic models of vaccine development, including prizes, tax incentives, and advance market commitments. A number of groups, including the Stop TB Partnership, the South African Tuberculosis Vaccine Initiative, and the Aeras Global TB Vaccine Foundation, are involved with research. Among these, the Aeras Global TB Vaccine Foundation received a gift of more than $280 million (US) from the Bill and Melinda Gates Foundation to develop and license an improved vaccine against tuberculosis for use in high burden countries.

A number of medications are being studied for multi drug resistant tuberculosis including: bedaquiline and delamanid. Bedaquiline received U.S. Food and Drug Administration (FDA) approval in late 2012. The safety and effectiveness of these new agents are still uncertain, because they are based on the results of a relatively small studies. However, existing data suggest that patients taking bedaquiline in addition to standard TB therapy are five times more likely to die than those without the new drug, which has resulted in medical journal articles raising health policy questions about why the FDA approved the drug and whether financial ties to the company making bedaquiline influenced physicians' support for its use

Other Animals

Mycobacteria infect many different animals, including birds, rodents, and reptiles. The subspecies *Mycobacterium tuberculosis*, though, is rarely present in wild animals. An effort to eradicate bovine tuberculosis caused by *Mycobacterium bovis* from the cattle and deer herds of New Zealand has been relatively successful. Efforts in Great Britain have been less successful.

As of 2015, tuberculosis appears to be widespread among captive elephants in the US. It is believed that the animals originally acquired the disease from humans, a process called reverse zoonosis. Because the disease can spread through the air to infect both humans and other animals, it is a public health concern affecting circuses and zoos.

Typhoid Fever

Typhoid fever, also known simply as typhoid, is a bacterial infection due to *Salmonella typhi* that causes symptoms which may vary from mild to severe and usually begin six to thirty days after exposure. Often there is a gradual onset of a high fever over several days. Weakness, abdominal pain, constipation, and headaches also commonly occur. Diarrhea is uncommon and vomiting is not usually severe. Some people develop a skin rash with rose colored spots. In severe cases there may be confusion. Without treatment symptoms may last weeks or months. Other people may carry the bacterium without being affected; however, they are still able to spread the disease to others. Typhoid fever is a type of enteric fever along with paratyphoid fever.

The cause is the bacterium *Salmonella typhi*, also known as *Salmonella enterica* serotype typhi, growing in the intestines and blood. Typhoid is spread by eating or drinking food or water contaminated with the feces of an infected person. Risk factors include poor sanitation and poor hygiene. Those who travel to the developing world are also at risk and only humans can be infected. Diagnosis is by either culturing the bacteria or detecting the bacterium's DNA in the blood, stool, or bone marrow. Culturing the bacterium can be difficult. Bone marrow testing is the most accurate. Symptoms are similar to that of many other infectious diseases. Typhus is a different disease.

A typhoid vaccine can prevent about 30% to 70% of cases during the first two years. The vaccine may have some effect for up to seven years. It is recommended for those at high risk or people traveling to areas where the disease is common. Other efforts to prevent the disease include providing clean drinking water, better sanitation, and better hand-washing. Until it has been confirmed that an individual's infection is cleared, the individual should not prepare food for others. Treatment of disease is with antibiotics such as azithromycin, fluoroquinolones or third generation cephalosporins. Resistance to these antibiotics has been developing, which has made treatment of the disease more difficult.

In 2013 there were 11 million new cases reported worldwide. The disease is most common in India. Children are most commonly affected. Rates of disease decreased in the developed world in the 1940s as a result of improved sanitation and use of antibiotics to treat the disease. Each year in the United States about 400 cases are reported and it is estimated that the disease occurs in about 6,000 people. In 2013 it resulted in about 161,000 deaths worldwide - down from 181,000 in 1990 (about 0.3% of the global total). The risk of death may be as high as 20% without treatment. With treatment it is between 1 and 4%. The name typhoid means "resembling typhus" due to the similarity in symptoms.

Signs and Symptoms

Classically, the course of untreated typhoid fever is divided into four distinct stages, each lasting about a week. Over the course of these stages, the patient becomes exhausted and emaciated.

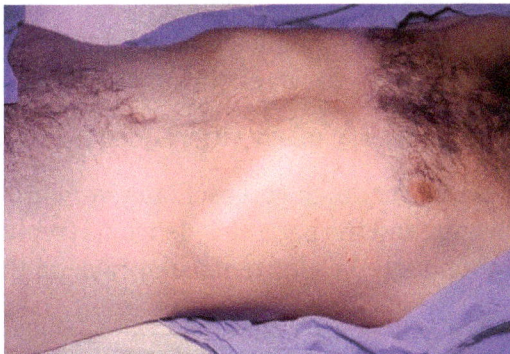

Rose spots on abdomen of a person with typhoid fever

- In the first week, the body temperature rises slowly, and fever fluctuations are seen with relative bradycardia (Faget sign), malaise, headache, and cough. A bloody nose (epistaxis) is seen in a quarter of cases, and abdominal pain is also possible. A decrease in the number of circulating white blood cells (leukopenia) occurs with eosinopenia and relative lymphocytosis; blood cultures are positive for *Salmonella typhi* or *S. paratyphi*. The Widal test is negative in the first week.

- In the second week, the person is often too tired to get up, with high fever in plateau around 40 °C (104 °F) and bradycardia (sphygmothermic dissociation or Faget sign), classically with a dicrotic pulse wave. Delirium is frequent, often calm, but sometimes agitated. This delirium gives to typhoid the nickname of "nervous fever". Rose spots appear on the lower chest and abdomen in around a third of patients. Rhonchi are heard in lung bases.

- The abdomen is distended and painful in the right lower quadrant, where borborygmi can be heard. Diarrhea can occur in this stage: six to eight stools in a day, green, comparable to pea soup, with a characteristic smell. However, constipation is also frequent. The spleen and liver are enlarged (hepatosplenomegaly) and tender, and liver transaminases are elevated. The Widal test is strongly positive, with antiO and antiH antibodies. Blood cultures are sometimes still positive at this stage.

- (The major symptom of this fever is that the fever usually rises in the afternoon up to the first and second week.)

- In the third week of typhoid fever, a number of complications can occur:

  - Intestinal haemorrhage due to bleeding in congested Peyer's patches; this can be very serious, but is usually not fatal.

  - Intestinal perforation in the distal ileum: this is a very serious complication and is frequently fatal. It may occur without alarming symptoms until septicaemia or diffuse peritonitis sets in.

  - Encephalitis

  - Respiratory diseases such as pneumonia and acute bronchitis

  - Neuropsychiatric symptoms (described as "muttering delirium" or "coma vigil"), with picking at bedclothes or imaginary objects.

  - Metastatic abscesses, cholecystitis, endocarditis, and osteitis

  - The fever is still very high and oscillates very little over 24 hours. Dehydration ensues, and the patient is delirious (typhoid state). One-third of affected individuals develop a macular rash on the trunk.

  - Platelet count goes down slowly and risk of bleeding rises.§

- By the end of third week, the fever starts subsiding

Cause

Transmission

The bacterium that causes typhoid fever may be spread through poor hygiene habits and public sanitation conditions, and sometimes also by flying insects feeding on feces. Public education campaigns encouraging people to wash their hands after defecating and before handling food are an important component in controlling spread of the disease. According to statistics from the United States Centers for Disease Control and Prevention (CDC), the chlorination of drinking water has led to dramatic decreases in the transmission of typhoid fever in the United States.

A 1939 conceptual illustration showing various ways that typhoid bacteria can contaminate a water well (center)

Bacteria

The cause is the bacterium *Salmonella typhi*, also known as *Salmonella enterica* serotype typhi.

There are two main types of Typhi namely the ST1 and ST2 based on MLST subtyping scheme, which are currently widespread globally.

Diagnosis

Diagnosis is made by any blood, bone marrow or stool cultures and with the Widal test (demonstration of antibodies against *Salmonella* antigens O-somatic and H-flagellar). In epidemics and less wealthy countries, after excluding malaria, dysentery, or pneumonia, a therapeutic trial time with chloramphenicol is generally undertaken while awaiting the results of the Widal test and cultures of the blood and stool.

The Widal test is time-consuming, and often, when a diagnosis is reached, it is too late to start an antibiotic regimen.

The term 'enteric fever' is a collective term that refers to severe typhoid and paratyphoid.

Prevention

Sanitation and hygiene are important to prevent typhoid. Typhoid does not affect animals other than humans. Typhoid can only spread in environments where human feces or urine are able to come into contact with food or drinking water. Careful food preparation and washing of hands are crucial to prevent typhoid. Industrialization, and in particular, the invention of the automobile, contributed greatly to the elimination of typhoid fever, as it eliminated the public health hazards associated with having horse manure in the public street which led to large number of flies.

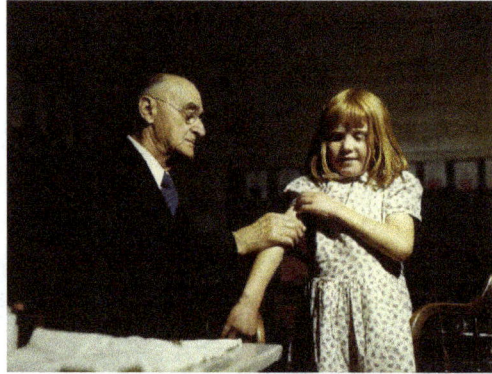

Doctor administering a typhoid vaccination at a school in San Augustine County, Texas, 1943

Two typhoid vaccines are licensed for use for the prevention of typhoid: the live, oral Ty21a vaccine (sold as Vivotif by Crucell Switzerland AG) and the injectable typhoid polysaccharide vaccine (sold as Typhim Vi by Sanofi Pasteur and Typherix by GlaxoSmithKline). Both are recommended for travellers to areas where typhoid is endemic. Boosters are recommended every five years for the oral vaccine and every two years for the injectable form. An older, killed-whole-cell vaccine is still used in countries where the newer preparations are not available, but this vaccine is no longer recommended for use because it has a higher rate of side effects (mainly pain and inflammation at the site of the injection).

Treatment

The rediscovery of oral rehydration therapy in the 1960s provided a simple way to prevent many of the deaths of diarrheal diseases in general.

Where resistance is uncommon, the treatment of choice is a fluoroquinolone such as ciprofloxacin. Otherwise, a third-generation cephalosporin such as ceftriaxone or cefotaxime is the first choice. Cefixime is a suitable oral alternative.

Typhoid fever, when properly treated, is not fatal in most cases. Antibiotics, such as ampicillin, chloramphenicol, trimethoprim-sulfamethoxazole, amoxicillin, and ciprofloxacin, have been commonly used to treat typhoid fever in microbiology. Treatment of the disease with antibiotics reduces the case-fatality rate to about 1%.

When untreated, typhoid fever persists for three weeks to a month. Death occurs in 10% to 30% of untreated cases. In some communities, however, case-fatality rates may reach as high as 47%.

Surgery

Surgery is usually indicated in cases of intestinal perforation. Most surgeons prefer simple closure of the perforation with drainage of the peritoneum. Small-bowel resection is indicated for patients with multiple perforations.

If antibiotic treatment fails to eradicate the hepatobiliary carriage, the gallbladder should be resected. Cholecystectomy is not always successful in eradicating the carrier state because of persisting hepatic infection.

Resistance

As resistance to ampicillin, chloramphenicol, trimethoprim-sulfamethoxazole, and streptomycin is now common, these agents have not been used as first–line treatment of typhoid fever for almost 20 years. Typhoid resistant to these agents is known as multidrug-resistant typhoid (MDR typhoid).

Ciprofloxacin resistance is an increasing problem, especially in the Indian subcontinent and Southeast Asia. Many centres are shifting from using ciprofloxacin as the first line for treating suspected typhoid originating in South America, India, Pakistan, Bangladesh, Thailand, or Vietnam. For these people, the recommended first-line treatment is ceftriaxone. Also, azithromycin has been suggested to be better at treating typhoid in resistant populations than both fluoroquinolone drugs and ceftriaxone. Azithromycin significantly reduces relapse rates compared with ceftriaxone.

A separate problem exists with laboratory testing for reduced susceptibility to ciprofloxacin: current recommendations are that isolates should be tested simultaneously against ciprofloxacin (CIP) and against nalidixic acid (NAL), and that isolates that are sensitive to both CIP and NAL should be reported as "sensitive to ciprofloxacin", but that isolates testing sensitive to CIP but not to NAL should be reported as "reduced sensitivity to ciprofloxacin". However, an analysis of 271 isolates showed that around 18% of isolates with a reduced susceptibility to ciprofloxacin (MIC 0.125–1.0 mg/l) would not be picked up by this method. How this problem can be solved is not certain, because most laboratories around the world (including the West) are dependent on disk testing and cannot test for MICs.

Epidemiology

In 2000, typhoid fever caused an estimated 21.7 million illnesses and 217,000 deaths. It occurs most often in children and young adults between 5 and 19 years old. In 2013 it resulted in about 161,000 deaths – down from 181,000 in 1990. Infants, children, and adolescents in south-central and Southeast Asia experience the greatest burden of illness. Outbreaks of typhoid fever are also frequently reported from sub-Saharan Africa and countries in Southeast Asia. Historically, in the pre-antibiotic era, the case fatality rate of typhoid fever was 10–20%. Today, with prompt treatment, it is less than 1%. However, about 3-5% of individuals who are infected will develop a chronic infection in the gall bladder. Since *S.* Typhi is human-restricted, these chronic carriers become the crucial reservoir, which can persist for decades for further spread of the disease, further complicating the identification and treatment of the disease. Lately, the study of Typhi associated with a large outbreak and a carrier at the genome level provides new insights into the pathogenesis of the pathogen.

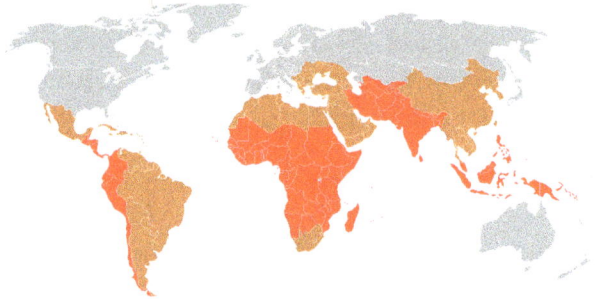

Incidence of typhoid fever
♦ Strongly endemic
♦ Endemic
♦ Sporadic cases

History

In 430 BC, a plague, which some believe to have been typhoid fever, killed one-third of the population of Athens, including their leader Pericles. Following this disaster, the balance of power shifted from Athens to Sparta, ending the Golden Age of Pericles that had marked Athenian dominance in the Greek ancient world. The ancient historian Thucydides also contracted the disease, but he survived to write about the plague. His writings are the primary source on this outbreak, and modern academics and medical scientists consider epidemic typhus the most likely cause. In 2006, a study detected DNA sequences similar to those of the bacterium responsible for typhoid fever in dental pulp extracted from a burial pit dated to the time of the outbreak.

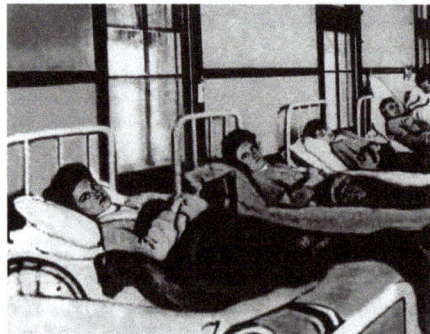

Mary Mallon ("Typhoid Mary") in a hospital bed (foreground): She was forcibly quarantined as a carrier of typhoid fever in 1907 for three years and then again from 1915 until her death in 1938.

The cause of the plague has long been disputed and other scientists have disputed the findings, citing serious methodologic flaws in the dental pulp-derived DNA study. The disease is most commonly transmitted through poor hygiene habits and public sanitation conditions; during the period in question related to Athens above, the whole population of Attica was besieged within the Long Walls and lived in tents.

Some historians believe that English colony of Jamestown, Virginia, died out from typhoid. Typhoid fever killed more than 6000 settlers in the New World between 1607 and 1624.

During the American Civil War, 81,360 Union soldiers died of typhoid or dysentery, far more than died of battle wounds. In the late 19th century, the typhoid fever mortality rate in Chicago averaged 65 per 100,000 people a year. The worst year was 1891, when the typhoid death rate was 174 per 100,000 people.

During the Spanish–American War, American troops were exposed to typhoid fever in stateside training camps and overseas, largely due to inadequate sanitation systems. The Surgeon General of the Army, George Miller Sternberg, suggested that the War Department create a Typhoid Fever Board. Major Walter Reed, Edward O. Shakespeare, and Victor C. Vaughan were appointed August 18, 1898, with Reed being designated the President of the Board. The Typhoid Board determined that during the war, more soldiers died from this disease than from yellow fever or from battle wounds. The Board promoted sanitary measures including latrine policy, disinfection, camp relocation, and water sterilization, but by far the most successful antityphoid method was vaccination, which became compulsory in June 1911 for all federal troops.

The most notorious carrier of typhoid fever, but by no means the most destructive, was Mary Mallon, also known as Typhoid Mary. In 1907, she became the first carrier in the United States to be identified and traced. She was a cook in New York who is closely associated with 53 cases and three deaths. Public health authorities told Mary to give up working as a cook or have her gall bladder removed, as she had a chronic infection that kept her active as a carrier of the disease. Mary quit her job, but returned later under a false name. She was detained and quarantined after another typhoid outbreak. She died of pneumonia after 26 years in quarantine.

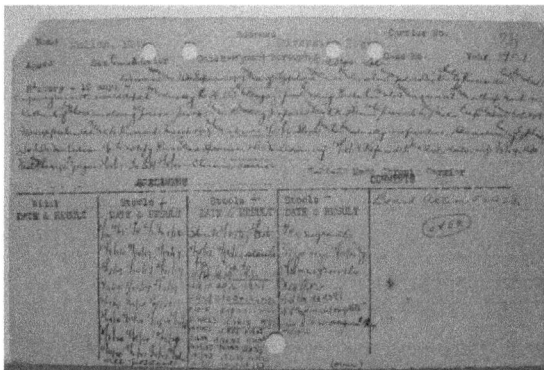

Original stool report for Mary Mallon, 1907.

Development of Vaccination

During the course of treatment of a typhoid outbreak in a local village in 1838, English country doctor William Budd realised the "poisons" involved in infectious diseases multiplied in the intestines of the sick, were present in their excretions, and could be transmitted to the healthy through their consumption of contaminated water. He proposed strict isolation or quarantine as a method for containing such outbreaks in the future. The medical and scientific communities did not identify the role of microorganisms in infectious disease until the work of Louis Pasteur.

Almroth Edward Wright developed the first effective typhoid vaccine.

In 1880, Karl Joseph Eberth described a bacillus that he suspected was the cause of typhoid. In 1884, pathologist Georg Theodor August Gaffky (1850–1918) confirmed Eberth's findings, and the organism was given names such as Eberth's bacillus, *Eberthella typhi*, and Gaffky-Eberth bacillus. Today, the bacillus that causes typhoid fever goes by the scientific name *Salmonella enterica enterica*, serovar Typhi.

The British bacteriologist Almroth Edward Wright first developed an effective typhoid vaccine at the Army Medical School in Netley, Hampshire. It was introduced in 1896 and used successfully by the British during the Boer War in South Africa. At that time, typhoid often killed more soldiers at war than were lost due to enemy combat. Wright further developed his vaccine at a newly opened research department at St Mary's Hospital Medical School in London from 1902, where he established a method for measuring protective substances (opsonin) in human blood.

Citing the example of the Second Boer War, during which many soldiers died from easily preventable diseases, Wright convinced the British Army that 10 million vaccines should be produced for the troops being sent to the Western Front, thereby saving up to half a million lives during World War I. The British Army was the only combatant at the outbreak of the war to have its troops fully immunized against the bacterium. For the first time, their casualties due to combat exceeded those from disease.

In 1909, Frederick F. Russell, a U.S. Army physician, adopted Wright's typhoid vaccine for use with the US Army, and two years later, his vaccination program became the first in which an entire army was immunized. It eliminated typhoid as a significant cause of morbidity and mortality in the U.S. military.

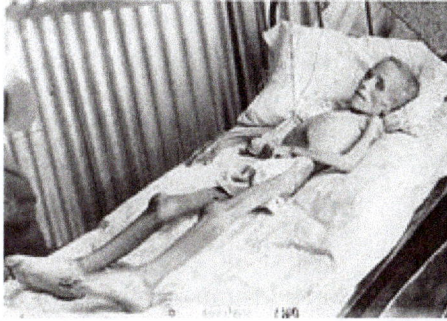

Lizzie van Zyl was a child inmate in a British-run concentration camp in South Africa who died from typhoid fever during the Boer War (1899–1902).

Most developed countries saw declining rates of typhoid fever throughout the first half of the 20th century due to vaccinations and advances in public sanitation and hygiene. In 1908, the chlorination of public drinking water was a significant step in the US in the control of typhoid fever. The first permanent disinfection of drinking water in the U.S. was made to the Jersey City, New Jersey, water supply. Credit for the decision to build the chlorination system has been given to John L. Leal. The chlorination facility was designed by George W. Fuller. In 1942 doctors introduced antibiotics in clinical practice, greatly reducing mortality. Today, the incidence of typhoid fever in developed countries is around five cases per million people per year.

A notable outbreak occurred in Aberdeen, Scotland, in 1964. This was due to contaminated tinned meat sold at the city's branch of the William Low chain of stores. No fatalities resulted.

In 2004–05 an outbreak in the Democratic Republic of Congo resulted in more than 42,000 cases and 214 deaths.

Names

The disease has been referred to by various names, often associated with symptoms, such as gastric fever, enteric fever, abdominal typhus, infantile remittant fever, slow fever, nervous fever, and pythogenic fever.

Bacterial Cold Water Disease

Bacterial cold water disease (BCWD) is a bacterial disease of salmonid fish. It is caused by *Flavobacterium psychrophilum* (previously classified in the genus *Cytophaga*), a gram-negative rod-shaped bacterium of the family Flavobacteriaceae. The disease typically occurs at temperatures below 13°C, and it can be seen in any area with water temperatures consistently below 15°C. Salmon are the most commonly affected species. This disease is not zoonotic.

Asymptomatic carrier fish and contaminated water provide reservoirs for disease. Transmission is mainly horizontal, but vertical transmission can also occur.

BCWD may be referred to by a number of other names including cold water disease, peduncle disease, fit rot, tail rot and rainbow trout fry mortality syndrome.

Diagnosis

Fish infected with typical BCWD have lesions on the skin and fins. Fins may appear dark, torn, split, ragged, frayed and may even be lost completely. Affected fish are often lethargic and stop feeding. Infection may spread systemically. Salmonid fish can also get a chronic form of BCWD following recovery from typical BCWD. It is characterised by erratic "corkscrew" swimming, blackened tails and spinal deformities.

The sweetfish which was infected with cold water disease

In rainbow trout fry syndrome, acute disease with high mortality rates occurs. Infected fish may show signs of lethargy, inappetance and exopthalmos before death.

A presumptive diagnosis can be made based on the history, clinical signs, pattern of mortality and water temperature, especially if there is a history of the disease in the area. The organism can be cultured for definitive diagnosis. Alternatively, histology should show periostitis, osteitis, meningitis and ganglioneuritis.

Treatment

Quaternary ammonium compounds can be added to the water of infected adult fish and fry. Alternatively, the antibiotic oxytetracycline can be given to adults, fry and broodstock. To prevent the disease, it is necessary to ensure water is pathogen-free and that water hardening is completed effectively for eggs.

Beet Vascular Necrosis

Beet vascular necrosis and rot is a soft rot disease caused by the bacterium *Pectobacterium carotovorum* subsp. *betavasculorum*, which has also been known as *Pectobacterium betavasculorum* and *Erwinia carotovora* subsp. *betavasculorum*. It was classified in the genus *Erwinia* until genetic evidence suggested that it belongs to its own group; however, the name Erwinia is still in use. As such, the disease is sometimes called Erwinia rot today. It is a very destructive disease that has been reported across the United States as well as in Egypt. Symptoms include wilting and black streaks on the leaves and petioles. It is usually not fatal to the plant, but in severe cases the beets will become hollowed and unmarketable. The bacteria is a generalist species which rots beets and other plants by secreting digestive enzymes that break down the cell wall and parenchyma tissues. The bacteria thrive in warm and wet conditions, but cannot survive long in fallow soil. However, it is able to persist for long periods of time in the rhizosphere of weeds and non-host crops. While it is difficult to eradicate, there are cultural practices that can be used to control the spread of the disease, such as avoiding injury to the plants and reducing or eliminating application of nitrogen fertilizer.

A table beet infected with *Pectobacterium carotovorum* subsp. *betavasculorum*. Note the rings of black vascular tissue colonized by the rotting bacteria.

Hosts

Fodder beets, sugar beets and fodder-sugar crosses are all susceptible to infection by *Pectobacterium carotovorum* subsp. *betavasculorum*. Today most beet cultivars are resistant to the pathogen, however, isolates vary geographically, and some cultivars of beets are only resistant to specific isolates of bacteria. For example, the cultivar USH11 demonstrates resistance to both Montana and California isolates, whereas Beta 4430 is highly susceptible to the Montana isolates but resistant to the California isolate. Other cultivars resistant to California isolates of *Pectobacterium caratovorum* subsp. *betavasculorum* include Beta 4776R, Beta 4430R and Beta 4035R, but HH50 has been found to be susceptible.

Breeding for resistance to other diseases such as beet yellows virus without also selecting for vascular necrosis resistance can leave cultivars susceptible to the pathogen. For example, the use of USH9A and H9B in California's San Joaquin valley is thought to have led to an epiphytotic (severe) outbreak of disease in the early 1970s. This was likely because of the limited gene pool used when selecting strongly for resistance to beet yellows virus. Further information on resistant cultivars can be found in the section Management.

In addition to beets, *Pectobacterium carotovara* subsp. *betavasculorum* can also infect tomato, potato, carrots, sweet potato, radish, sunflower, artichokes, squash, cucumber and chrysanthemum. Other subspecies of *Pectobacterium carotovora* can also be pathogenic to beets. *Erwinia carotovara* subsp. *atroseptica* is a bacterial soft rot pathogen that is responsible for the disease Blackleg of Potato (*Solanum tuberosum*), and variants of this bacterium can cause root rot in sugarbeets,. This subspecies also has a wide host-range. *Erwinia carotovora* var. *atroseptica* has been detected in the rhizosphere of native vegetation and on weed species such as *Lupinus blumerii* and *Amaranthus palmeri* (pigweed). It is thought that the source of inoculums survives on these non-host plants in areas in which it is endemic as well as in the rhizosphere of other crops such as wheat and corn

Symptoms

Symptoms can be found on both beet roots and foliage, although foliar symptoms are not always present. If present, foliar symptoms include dark streaking along petioles and viscous froth deposits on the crown which are a by-product of bacterial metabolism. Petioles can also become necrotic and demonstrate vascular necrosis. When roots become severely affected, wilting also occurs. Below ground symptoms include both soft and dry root rot. Affected vascular bundles in roots become necrotic and brown, and tissue adjacent to necrosis becomes pink upon air contact. The plants that do not die completely may have rotted-out, cavernous roots.

Table beet stem infected with *Pectobacterium carotovorum* subsp. *betavasculorum*. Note entry through a wound.

Various pathogens can cause root rot in beets; however the black streaking on petioles and necrotic vascular bundles in roots and adjacent pink tissue help to distinguish this disease from others such as Fusarium Yellows. Additionally, sampling from the rhizosphere of infected plants and serological tests can confirm the presence of *Erwinia caratovora* subs.

Disease Cycle

Pectobacterium carotovorum subsp. *betavasculorum* is a gram negative, rod bacteria with peritichous flagella. For it to enter sugar beet, and thus cause infection, it is essential that there is an injury to the leaves, petioles or crown. Infection will often start at the crown and then move down into the root, and can occur at any point in the growing season if environmental conditions are favorable. Once the bacteria enters the plant, it will invade the vascular tissue and cause symptoms by producing plant cell wall degrading enzymes, like pectinases, polygalactronases, and celluases. This results in discolored or necrotic vascular tissue in the root, and the tissue bordering the vascular bundles will turn reddish upon contact with air. Following the infection of the vascular tissue, the bacteria reproduce as long as food resources are available, and the root begins to rot. There is significant variability in the type of rot – it can range from a dry rot to soft and wet rot – because of the multitude of additional microorganisms that may colonize the damaged tissue

Upon death of the sugar beet, or harvest of the field, the pathogen appears to survive in select living plant tissue like beet roots, or volunteer beets. However, it does not appear to survive in sugar beet seeds, or live in the soil after harvest. It is also possible for the pathogen to infect injured carrots, potato, sweet potato, tomato, radish, sunflower, artichokes, squash, cucumber and chrysanthemums; however, since those are often planted in the same season as sugar beets, they are not likely to be overwintering hosts.

Environment

Injury to the leaves, petioles or crown is mandatory for the pathogen to gain entry to the host tissue. Accordingly, hail damage is correlated with a higher degree of disease outbreak. Young plants (less than eight weeks old) are also considered to be more prone to infection

Temperature and availability of moisture are key factors in determining the rate of disease development. Warm temperatures, 25-30 °C, promote rapid disease development., and can result in acute symptoms. Symptoms are also reported to appear at temperatures as low as 18 °C, but disease development is slowed; below that temperature, infections do not develop. Excessive water also promotes disease development by providing a more optimal environment for the pathogen, and has been shown to be a key factor in augmenting disease outbreak in fields with sprinkler irrigation

Agricultural

The degree of nitrogen fertilization is highly correlated to robust disease development: it has been shown that sugar beets supplied with excessive or adequate nitrogen are more diseased than sugarbeets with sub-optimal nitrogen levels. This is a paradox for farmers because, while increased nitrogen fertilization does increase sugar yield in non-infected sugarbeets, it also increases the severity of the disease if infection takes place. Thus, depending on the severity of infection, yield may go down with increased fertilizer use.

The spacing between plants also impacts the degree of infection: greater in-row spacing results in more diseased roots. This may be due to the fact that greater spacing promotes faster growth, and hence greater probability of cracks in the crown, or because of the increased amount of nitrogen available per plant.

Since the pathogen has multiple hosts, it is important for farmers to be wary of other plants in the surrounding area. It is possible for the pathogen to survive in weedy hosts, and can infect injured carrots, potato, sweet potato, tomato, radish, squash, and cucumber. Hence, the presence of these plants may increase the supply of inoculum.

Laboratory

If the pathogen is cultured in a lab, it can grow on Miller and Schroth media, can use sucrose to make reducing sugars, and can use either lactose, methyl alpha-glucoside, inulin or raffinose to make acids. It is also capable of surviving in culture medium sodium levels of up to 7–9%, and in temperatures as high as 39 °C.

Management

Since the bacteria cannot survive in seeds, the best way to prevent the disease is to ensure that vegetatively propagated plant material are clean of infection, such that the bacterium does not enter the soil. However, if the bacteria is already present, there are some methods that can be used to lessen the infection.

Cultural Practices

Because the bacteria readily enter the plant through wounds, management practices that decrease injury to the plants are important to control the spread of the disease. Cultivation is not recommended, as the machinery can become contaminated and physically spread the bacteria around the soil. Accidental leaf tearing or root scarring can also occur depending on the size of the crop, allowing the bacteria to enter more individual plants. If hilling the beets, great care must be taken to avoid getting soil into the crown, because the pathogen is soil-borne and this could expose the plant to more bacteria, thus increasing the risk of infection.

While most bacteria are motile and can swim, they cannot move very far due to their small size. However, they can be carried along by water, and a significant movement of *Pectobacterium* can be attributed to being carried downstream from irrigation and rainwater. To control the spread of the disease, limiting irrigation is another strategy. The bacteria also flourishes in wet conditions, so limiting excess water can control both the spread and severity of the disease.

Increased in-row spacing also causes more severe disease. In an infected field, yield decreased linearly when spacing was greater than 15 cm (6 in), so a spacing of 6 inches or less is recommended.

The bacteria can also utilize nitrogen fertilizer to accelerate their growth, thus limiting or eliminating the amount of nitrogen fertilizer applied will lessen the disease severity. For example, when fertilizer was applied to an infected field the infection rate per root increased from 11% (with no added nitrogen) to 36% (with 336 kg nitrogen/hectare), and sugar yields decreased.

Resistance

The bacteria can survive in the rhizosphere of other crops such as tomato, carrots, sweet potato, radish, and squash as well as weed plants like lupin and pigweed, so it is very hard to get rid of it completely. When it is known that the bacterium is present in the soil, planting resistant varieties can be the best defense against the disease. Many available beet cultivars are resistant to *Pectobacterium carotovorum* subsp. *betavasculorum*, and some examples are provided in the corresponding table. A comprehensive list is maintained by the USDA on the Germplasm Resources Information Network. Even though some genes associated with root defense response have been identified, the specific mechanism of resistance is unknown, and it is currently being researched.

Biological Control

Some bacteriophages, viruses that infect bacteria, have been used as effective controls of bacterial diseases in laboratory experiments. This relatively new technology is a promising control method that is currently being researched. Bacteriophages are extremely host-specific, which makes them environmentally sound as they will not destroy other, beneficial soil microorganisms. Some bacteriophages identified as effective controls of *Pectobacterium carotovorum* subsp. *betavasculorum* are the strains ΦEcc2 ΦEcc3 ΦEcc9 ΦEcc14. When mixed with a fertilizer and applied to inoculated calla lily bulbs in a greenhouse, they reduced diseased tissue by 40 to 70%. ΦEcc3 appeared to be the most effective, reducing the percent of diseased plants from 30 to 5% in one trial, to 50 to 15% in a second trial. They have also been used successfully to reduce rotting in lettuce caused by *Pectobacterium carotovorum* subsp. *carotovorum*, a different bacterial species closely related to the one that causes beet vascular necrosis.

While it is more difficult to apply bacteriophages in a field setting, it is not impossible, and laboratory and greenhouse trials are showing bacteriophages to potentially be a very effective control mechanism. However, there are a few obstacles to surmount before field trials can begin. A large problem is that they are damaged by UV light, so applying the phage mixture during the evening will help promote its viability. Also, providing the phages with susceptible non-pathogenic bacteria to replicate with can ensure there is adequate persistence until the bacteriophages can spread to the targeted bacteria. The bacteriophages are unable to kill all the bacteria, because they need a dense population of bacteria in order to effectively infect and spread, so while the phages were able to decrease the number of diseased plants by up to 35%, around 2,000 Colony Forming Units per milliliter (an estimate of living bacteria cells) were able to survive the treatment. Lastly, the use of these bacteriophages places strong selection on the host bacteria, which causes a high probability of developing resistance to the attacking bacteriophage. Thus it is recommended that multiple strains of the bacteriophage be used in each application so the bacteria do not have a chance to develop resistance to any one strain.

Importance

The disease was first identified in the western states of, California, Washington, Texas, Arizona and Idaho in the 1970s and initially led to substantial yield losses in those areas. *Erwinia caratovara* subsp *betavascularum* was not discovered in Montana until 1998. When it first appeared, beet vascular necrosis caused individual farm yield loss ranging from 5–70% in Montana's Bighorn Valley. Today, yield losses from the disease are generally infrequent and patchy as most producers plant resistant varieties. Infection rate is generally low if resistant cultivars are chosen; however, warmer and wetter conditions can lead to higher than normal instance of disease

If infection does occur, bacterial root rots can not only cause economic losses in the field, but also can in storage and processing as well. In processing plants, rotten roots complicate slicing and the bacterially-produced slime can clog filters. This is especially problematic with late-infected beets which are generally harvested and processed along with healthy beets. The disease can also lower sugar-content which greatly reduces the quality

References

- Corey, Ralph (1990). "Ch. 39: Hemoptysis". In Walker HK, Hall WD, Hurst JW. Clinical Methods: The History, Physical, and Laboratory Examinations (3rd ed.). Boston: Butterworths. ISBN 0-409-90077-X.

- McMenamin, Dorothy (2011). Leprosy and stigma in the South Pacific : a region-by-region history with first person accounts. Jefferson, N.C.: McFarland. p. 17. ISBN 978-0-7864-6323-7.

- Ryan, Kenneth J.; Ray, C. George, eds. (2004). Sherris Medical Microbiology (4th ed.). McGraw Hill. pp. 451–3. ISBN 0-8385-8529-9. OCLC 61405904.

- McMurray DN (1996). "Mycobacteria and Nocardia". In Baron S; et al. Baron's Medical Microbiology (4th ed.). Univ of Texas Medical Branch. ISBN 0-9631172-1-1. OCLC 33838234.

- Andrew Baum; et al. (1997). Cambridge handbook of psychology, health and medicine. Cambridge, Angleterre: Cambridge University Press. p. 521. ISBN 978-0-521-43686-1.

- Hamilton, Bernard (2000). The leper king and his heirs: Baldwin IV and the Crusader Kingdom of Jerusalem. Cambridge, UK: Cambridge University Press. ISBN 0-521-64187-X.

- Bryant A (1995). Sekigahara 1600: The Final Struggle for Power (Campaign Series, 40). Osprey Publishing (UK). ISBN 1-85532-395-8. Retrieved 2010-02-28.

- Dolin, [edited by] Gerald L. Mandell, John E. Bennett, Raphael (2010). Mandell, Douglas, and

Bennett's principles and practice of infectious diseases (7th ed.). Philadelphia, PA: Churchill Livingstone/Elsevier. pp. Chapter 250. ISBN 978-0-443-06839-3.

- Organization, World Health (2008). Implementing the WHO Stop TB Strategy: a handbook for national TB control programmes. Geneva: World Health Organization. p. 179. ISBN 9789241546676.

- Gibson, Peter G. (ed.); Abramson, Michael (ed.); Wood-Baker, Richard (ed.); Volmink, Jimmy (ed.); Hensley, Michael (ed.); Costabel, Ulrich (ed.) (2005). Evidence-Based Respiratory Medicine (1st ed.). BMJ Books. p. 321. ISBN 978-0-7279-1605-1.

- Kabra, [edited by] Vimlesh Seth, S.K. (2006). Essentials of tuberculosis in children (3rd ed.). New Delhi: Jaypee Bros. Medical Publishers. p. 249. ISBN 978-81-8061-709-6.

- Ghosh, editors-in-chief, Thomas M. Habermann, Amit K. (2008). Mayo Clinic internal medicine: concise textbook. Rochester, MN: Mayo Clinic Scientific Press. p. 789. ISBN 978-1-4200-6749-1.

Permissions

Index

www.ingramcontent.com/pod-product-compliance
Lightning Source LLC
Chambersburg PA
CBHW061949190326
41458CB00009B/2828